GILLINGHAM'S ROYAL FOREST

The Medieval Centuries

JOHN PORTER

Gillingham Museum

First published 2013

The Gillingham Museum and Local History Society
Chantry Fields
Gillingham
Dorset SP8 4UA

www.gillinghammuseum.co.uk

© John Porter 2013

The right of John Porter to be identified as the Author of this work has been asserted in accordance with the Copyrights, Designs and Patents Act 1988.

All rights reserved. No part of this book may be reprinted or reproduced or utilised in any form or by any electronic, mechanical or other means, now known or hereafter invented, including photocopying and recording, or in any information storage or retrieval system, without permission in writing from the Publishers.

British Library Cataloguing in Publication Data.
A catalogue record for this book is available from the British Library.

ISBN 978-0-9927063-0-2

Typesetting and design by the author

Printed and bound by CPI Group (UK) Ltd, Croydon, CR0 4YY

Contents

Preface and Acknowledgements		3
1	Part of Selwood Forest	5
2	Of Bounds and Perambulations	11
3	The Royal Visits	17
4	King's Court: Palace and Park	26
5	Forest Guardians and Justice	40
6	'The Master of Game'	55
7	The Green Cover	67
8	A Well-Populated Forest	81
9	After King's Court	90
Glossary of Forest Terms		98
Notes and References		100
Bibliography of Works Used		107
Index		109

Preface

On the edge of the town of Gillingham, a small rectilinear earthwork in the corner of a field marks one of the most important historic sites of north Dorset. The site is not well known, a possible reason being that there is no signpost from the nearby main road to attract the attention of the visitor. The site itself, known as King's Court, is in the parish of Motcombe, and historically can claim to belong to both places.

King's Court in the Middle Ages was the royal residence or 'palace' of the Forest of Gillingham. King's Court and its forest have long ago faded into the rural scenery of north Dorset, but at the time of King John featured prominently in the journeys made by itinerant monarchs across the south of England. Here the Angevin monarchs could relax and draw breath after a long journey. In the adjoining park and forest, venison and other game were hunted for the royal sport and larder. From the forest, timbers were supplied to royal buildings and religious houses for many miles around.

Although the Forest of Gillingham was one of England's smaller royal forests, it is well documented in the many records of the royal household and departments, notably the Pipe Rolls, Close Rolls, and Liberate Rolls. It is fortunate that most of these were translated many years ago, and are readily available in the archives of the Gillingham Museum. Another illuminating source for the history of the medieval forest has been the much later forest map of 1624, to be found in the Dorset History Centre, which reveals the state of the forest in earlier centuries.

While these sources may show officialdom at work in the forest, they cannot tell the full story. The practices and customs of the royal hunt in Gillingham, with its distinctive procedures and rituals, have gone largely unrecorded, but there is no reason to suppose that hunting practices in Gillingham were any different from those of other, larger royal forests. Outside the royal and official presence, which was occasional and infrequent, the forest was left in the hands of its keepers, foresters, carpenters, labourers, and copyhold tenants, people whose names and daily lives are for the most part not recoverable.

The story of Gillingham Forest does not end with the medieval centuries. In the seventeenth century its disafforestation produced an episode of discontent and disorder which lasted for more than twenty years. Since these later events are distinctive in themselves, they will be related in a later publication.

In producing this account, I am grateful to the Gillingham Local History Society and Museum for their agreement to publish this work and for their continued encouragement. I am particularly indebted to Bill Shreeves for the use of his earlier work on the forest, much of which has appeared in recent issues of the *Journal of the Gillingham Local History Society*; and to David Lloyd and Sam Woodcock for their time spent with me in reviewing the drafts of the chapters. Further thanks are due to Caroline Johnston and Laura Porter for their comments on artwork and layout; to Laura for her expert sketching of several of the text figures; and to my wife Margaret for her support and attention to the finer points of grammar and punctuation. The drawn maps and photographs are my own. Time will reveal the errors and misinterpretations which are mine alone.

John Porter
September 2013

Acknowledgements are made to Gillingham Museum for Plate 1 and Fig. 20; and to the British Library for permission to use sketches based on the Luttrell Psalter for Figs. 3, 13, and 18 (BL ADD. 42130 f 206v, f68r and f163r).

CHAPTER 1

PART OF SELWOOD FOREST

In June of 1204 King John arrived for a stay of four nights in the Dorset town of Gillingham. He came with a large entourage which included his court, his family, officers of state, household knights, nobles with their own followers, and a party of huntsmen. His stay was notable for the feast he provided for the poor of the town. The venue was a hunting lodge on the edge of Gillingham, which had been newly rebuilt to a standard fit to receive a royal visit. The residence, which was to become known as 'King's Court', was to be visited several more times by John during his reign. King's Court was set within a royal deer park, which itself was on the edge of the extensive forest of Gillingham. It was the opportunity to spend some time hunting in the forest which was the attraction for the king's visit.

The term 'forest' at that time had a meaning different from that of the present day. Forests were areas under royal jurisdiction, set aside for the exclusive use of the king, and subject to especially severe laws directed at enforcing the king's rights. Like many forests, Gillingham was heavily wooded, but there were other royal forests in England consisting largely of heathland or moorland.

Because of parish boundary changes in later centuries, King's Court and the greater part of the former Gillingham forest now lie in the parish of Motcombe, only a smaller part of the forest being within the present day parish of Gillingham. Today, the old forest is an area of farms, hedged meadows and dairy pastures, much like its neighbouring areas of north Dorset. Nevertheless, there are a few features in the present day landscape which suggest that the area may have had a more distinctive history. The visitor perusing the map of the countryside around Gillingham and Motcombe might be attracted by the reference to the site of the old royal residence, marked today as 'King's Court Palace'. Nearby on the same map is a *Park Pale*, a type of fence or hedge usually associated with medieval deer parks. In the adjoining countryside names like Forest Lodge, Forest Farm, Forest Deer, and Huntingford suggest a past associated with woodlands and hunting.

Gillingham's Royal Forest: The Medieval Centuries

A journey to tell the story of King's Court and its forest leads back to the later medieval centuries, when Gillingham was one of a number of royal forests extending across the south of England. These forests had their origins in a very much larger forest or 'wildwood' which had once stretched across the Wessex countryside, and which was to become known to antiquarians and early historians as *Sealwudu* or Selwoodshire, the shire of the 'sallow wood'. This densely wooded tract lay mostly on the clay vales and limestone ridges which cut across Wessex from western Dorset to northern Wiltshire (Figure 1), reaching its highest point on the hill now occupied by Alfred's Tower. From this well known, very much later, landmark, the view takes in the extent of the great forest in all directions.

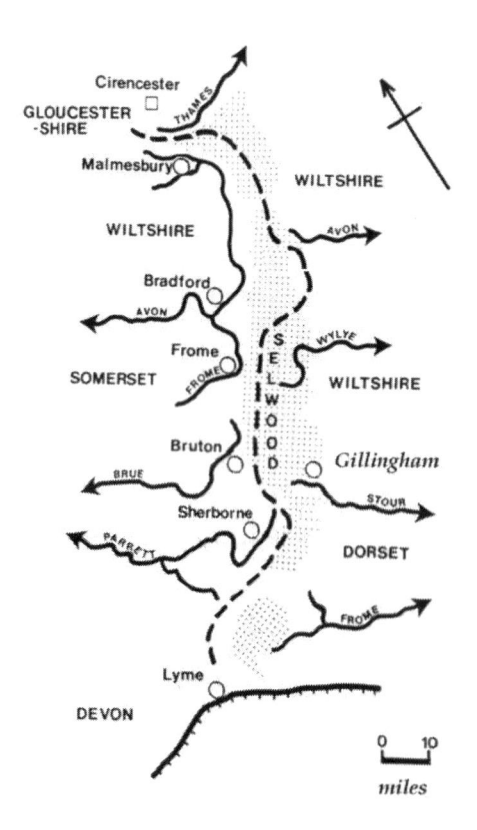

1 The ancient forest of Selwood or Selwoodshire. In pre-Conquest times this forest covered much of Somerset, Wiltshire, and Dorset (after Barker).

Long after the great forest had been much reduced in size and divided between later owners, its memory and legacy lived on. In the reign of Henry VIII, the Tudor traveller, John Leland, used the name Selwood and estimated it to extend thirty miles one way towards Warminster and ten miles the other way as far as Shaftesbury. Thomas Gerard, the Dorset topographer who lived a century later, noted that Gillingham could still be remembered as 'heretofore part of Selwood forest, only distinguished by names of places.'[2]

With the spread of royal estates around the Somerset area in Saxon times, Selwood would have been well used for hunting by the pre-Conquest kings of Wessex. King Alfred excelled in the art of hunting, while it is said that Edward the Confessor went hunting and hawking every day after his devotions. Reference to the hunting practices of the time appears in the *Dialogues (or Colloquy) of Aelfric,* believed to have been written around the turn of the eleventh century. Aelfric was a monk who spent many years at Cerne Abbey and later at Eynesham in Oxfordshire, and would have known the Wessex forests well. The *Dialogues* takes the form of 'question and answer' discussion between teacher and rural inhabitants, and gives a vivid picture of life in the Saxon countryside. In one section, a king's huntsman relates to Aelfric that hunting is his one and only skill, one which he performs with nets:

> I take my nets with me and set them in a suitable place, and set my hounds to pursue the beasts so that they reach the nets unexpectedly and are ensnared. Then, while they are still trapped in the nets, I cut their throats.

He then goes on to relate how he took harts, boars, roe deer and hares, with hounds, nets and spears. Aelfric's huntsman renders all his kills to the king, who in turn rewards him with food, clothes, horses, and armour.[3]

The extent to which the Saxon kings were specifically acquainted with the forest around Gillingham is more conjectural. William of Malmesbury claimed that Edward the Confessor had been chosen king while at Gillingham in 1042, and although now historically discounted, the idea may reflect something more than a passing acquaintance between the king and the town or forest. Gillingham is not described as a forest in Domesday Book, but the folios nevertheless provide substantial evidence of the value of Gillingham to Edward. Among the smaller estates listed as belonging to the royal manor is one belonging to Edward the huntsman, who had half of a virgate or ploughland, around 20 acres. The setting aside of a small estate for the maintenance of a huntsman suggests something about the importance of this role to the king. It is tempting to equate this holding with one which emerges in the forest records of some two centuries later. This is the tenement or bailiwick of the forester of fee, the king's chief keeper of the forest, who in the thirteenth century was still there to ensure that the forest met the king's needs in timber and venison.[4]

The coming of the Normans brought a new, more rigorous form of control over the forests. During the following century, the newer, more severe laws of the royal forest were ruthlessly imposed by the decrees of the Conqueror's successors, and further areas subjected to the forest laws.[6] In the south of England the most extensive area to be afforested by the Normans was the New Forest, but forest law was applied to many parts of Hampshire, Wiltshire, Dorset, Somerset, and Devon. In Dorset, the new royal forests were those of Gillingham, Blackmore, Bere, and Powerstock. The east of the county also included the great Cranborne Chase, which although never designated as a royal forest, came to enjoy the same laws and institutions as forests.[7] By the thirteenth century, forests covered so much of this part of Wessex that it was possible to ride most of the way from Southampton to Bath without leaving the jurisdiction of the forest law.

Throughout the later medieval centuries the extent of royal forests was a frequent cause of conflict between the king and his barons, the one party seeking to maintain the exclusive nature of the forest, and the other regularly seeking to have it cut back and diminished. In Stephen's reign the forest laws fell into abeyance because of the weakness of the king, but Henry II reinforced them throughout the kingdom with vigour and determination. In 1184 the king issued the Assize of Woodstock, a redefinition of forest law which strengthened his hold over the forests by requiring the appointment of officials known as verderers and agisters.[8] Even a small forest like Gillingham could not escape the scrutiny of the Chief Justice of the Forest, the king's chief forest official. He succeeded in extending its boundaries southwards to include Stour Provost, Todber, and Marnhull, taking in a tract of country almost reaching to Sturminster Newton, trebling the area under the forest laws. However, the extension could not outlive the troubles of John's reign, when the barons were able to compel the king to cede to them many rights and relax the forest laws. In 1217, following the issuing of the Charter of the Forest, all lands afforested by Henry II or his sons were to be disafforested. The forest of Gillingham in 1225 was cut back to its earlier limits, within which it subsequently remained.[8]

Part of Selwood Forest

During these centuries Gillingham forest was part of a wider royal jurisdiction, known as the manor or liberty of Gillingham. The royal forest lay on the eastern side of the manor, and while considered to be part of the manor, came to have its own separate administration in which the forest laws were imposed over the normal manorial laws. The village or town of Gillingham lay just across the Shreen from the forest, and by the thirteenth century had expanded across the river into a new settlement known as Newbury, which lay entirely within the forest boundary. From the time of Edward I the manor was always given to the queen as a jointure, and the forest is frequently referred to as 'the Queen's forest'.[9]

Gillingham forest was small by comparison with some royal forests, occupying 13.7 square miles (35.4 square km) or around 8,800 acres, and measuring no more than five miles from north to south. The forest was essentially the basin of the Lodden and the eastern bank of the Shreen, tributaries of the Stour which meet the main stream just outside the old town centre of Gillingham (Figure 4). Today it consists mostly of meadows and pastures with many hedges and occasional coppices, largely the product of post-medieval enclosure. In the early Middle Ages the landscape would have been entirely unenclosed, with tracts of dense oak woodland growing profusely on the low, undulating surfaces of the Kimmeridge Clay. Eastwards the clay gives way to the open, drier grounds of the Greensand around Kingsettle, Little Down, and the Wiltshire boundary, with a greater variety of trees.

Such an area, with its plentiful cover for deer, wild boar, birds, and wildfowl, could provide a king with space to relax, away from the constant demands of the court. The court at this time was not fixed in London or Westminster, but was continuously on the move around the country, stopping at many places to carry out essential administration, hear petitions, or dispense justice. A break from travel in one of the many royal forests gave the king the opportunity to gather his hunting party and get away for a few hours from the constant attentions of royal officials, courtiers, justices, petitioners, and all who made up the retinue of the travelling court. Gillingham forest was reasonably accessible to royal and official parties following the few well-trodden routes across the southern counties. From another royal lodge or palace at Clarendon, east of the Avon in Wiltshire, travellers could cross the Avon south of the later city of

New Sarum or Salisbury, and join the ridgeway road to the abbey town of Shaftesbury. From there the route continued into the forest along the ancient road which leads to the castle and abbey at Sherborne. From this road, it was only a small diversion off to Gillingham's royal residence or hunting lodge, standing by the River Lodden just outside the town.

Around the forest, there are a few places where the landscape and environment known to the royal party can still be envisaged. From Castle Hill in Shaftesbury the view takes in the forest beyond Motcombe and Gillingham towards Stourhead and Alfred's Tower, a high point of the old forest of Selwood. The Gillingham forest took in the village of Enmore Green at the foot of Castle Hill, and from here towards the horizon the medieval landscape would have been of largely continuous woods broken by clearings and occasional fields. In the foreground the wooded areas around Motcombe Park (now Port Regis), although of much later date, may suggest something of the appearance of the medieval forest.

On the north side of Shaftesbury, the old forest is overlooked by the steep wooded slopes of The Cliffe and Kingsettle woods. These slopes have been occupied continuously by forest since ancient times. Duncliffe Hill, on the south side of the former forest, was a prominent boundary point rising steeply out of the Blackmore Vale and visible for miles around. From its summit, wooded in recent times, there are fine views of the forest through breaks in the trees.

As the forest evolved through Norman and Angevin times, it became divided for management into 'walks' or areas of woodland grazings, areas where the deer could be allowed to run free, and also spaces where tenants of the manor with rights of common were allowed to pasture their animals. These were known as the Lawn Walk, north of the Lodden; the Clear Walk, north and east of the park towards Motcombe; and the Woods End Walk, in the south and east of the forest towards Shaftesbury. As the medieval centuries progressed the walks were to diminish in extent as the result of new enclosures and settlement, but even in the sixteenth century the forest could still present a largely wooded appearance.

CHAPTER 2

OF BOUNDS AND PERAMBULATIONS

The extent and boundaries of Gillingham forest were known to the king and the royal officers through the carrying out of perambulations. Medieval perambulations were detailed surveys of the boundaries of lordships, manors, or forests, carried out hedge by hedge and in a circuit from one landmark to another. The purpose of the survey was not merely to define the boundary of the lordship or estate, but also to determine the rights of the lord over the ground. In royal forests it was necessary to ensure that the lord received the full dues from any human acts which took place within the forest bounds, for example fines from deer illegally killed. The Tudor lawyer, John Manwood, in his *Treatise of the Forest Laws*, wrote

> Forests must be meered and bounded with unmoveable marks, meers, and boundaries, either known by matter of record or by prescription ... these marks, meers, and boundaries are unremoveable because the king hath an interest in them.[1]

The detail collected from forest perambulations could contribute to maximising the fees and incomes due to the king from forest trespasses, or might be used in enquiries into rights being claimed in the forest for which no legal entitlement could be produced. Perambulations therefore formed a key part of the king's endeavours to enforce his forest laws, and the instruction to carry out a perambulation often indicates a determination to revive royal rights which had been allowed to lapse. Perambulations of the forests were an ingredient in the struggles between the king and his barons, as each side endeavoured to use the forest boundaries to define the extent of royal authority to its own advantage.

Perambulations were usually carried out by groups of sworn jurors, which might include some local knights and officials such as keepers or foresters, but also included knights and gentry appointed by the king. The survey would take the jury along streams and hedges, and between bridges, gates, prominent trees, and boundary stones. As they moved on, the jurors would look for answers to questions such as: How far in this direction does the boundary extend? Who has the rights to the land beyond the boundary? What is the name of this landmark? Has it always been known by this name, and if not, by what name was it previously known? [2]

The earliest known of the Gillingham forest perambulations or circuits dates from 1225. In 1299 Edward I, under pressure from his barons to relax the forest laws, ordered perambulations of many of his manors and forests, including Gillingham. This has a different starting point and a number of names are different, but the bounds are almost the same as in 1225. A similar exercise, carried out in 1568 at the behest of Elizabeth I, confirmed the 1300 survey with changes in names. In 1605 a perambulation was made of the whole manor, so that only a part of the boundaries described relate to the forest. This was reviewed and largely repeated in 1635 and 1638. The last known perambulation, of both manor and forest, took place around 1650, some years after the forest had become private property, but when the new landowners still needed evidence of the old forest boundaries.[3]

The earliest perambulation, of 1225, formed part of the Cerne Cartulary, a collection of charters and documents relating to the lands and properties of Cerne Abbey. The only known copy of this was found bound inside the *Book of Cerne*, an ancient collection of Latin gospels now preserved in the Cambridge University Library. The perambulation arose out of disputes between the regents of the king (Henry III) and the barons, in which the regents had found themselves in a weak position. They had been forced to concede a series of perambulations with the objective of reducing the areas under royal forest. The survey was led by Hugh de Neville, (the king's chief justice of all his forests); Brian de Insula (also known as de Lisle, or of the Isles); Henry of Cerne, William of Morville, and John of Lancelevere (justices). There follows a list of fourteen sworn knights. The jurors making the perambulation begin by stating that Alan de Nevill had afforested all the hilly parts of Dorset after the coronation of Henry II, 'and that these ought not to be forests.'

The 1225 survey differs in bounds from the later perambulations. On the western side of the forest near the town of Gillingham the bounds take a different path from those of the later surveys. The effect is to exclude from the forest a tract lying between the

Shaftesbury road and the Stour, extending across to the north of Madjeston. The alteration did not endure, and the area had been restored to the forest by the time of the 1300 survey. The excluded area seems to be that of the manor of Ham, granted to the priory of Montacute in 1158, making it likely that the terms of the grant included freedom from the forest laws.[5] The circumstances leading to the later restoration to the forest are not known.

Huntingford to Kingsettle

The boundary points of the 1225 perambulation can be followed on the map of the forest (Figure 4), the first part of the perambulation reading as follows:

> Huntingford, and so always alongside the stream along the hill to the marches of Dorset and Wiltshire. And so proceeding as the march between the said two counties stretches as far as Legh; from Legh always by the boundaries of the counties as far as Pimperlegh; from Pimperlegh as far as Horr Apeldure always along the valley.[6]

There seems to be no particular reason for starting the circuit at Huntingford, although the name suggests that the ford across the Shreen here would have been a point well known to huntsmen (Plate 4). A bridge is not recorded until the 1300 survey.[7] From here the boundary followed the Shreen upstream for a short distance before joining the county boundary between Dorset and Wiltshire, which it followed for several miles. The perambulation now introduces some names which have since been forgotten. The first stretch along the county boundary was over *La Leghe* or the Leighe (wooded glade or clearing), known by Tudor times as White Hill, a belt of higher ground between the manors of Gillingham and Mere. The 1300 survey makes reference here to the wood of Horsington on the Mere side of the boundary, but by 1568 it had become only 'of old time a wood, which is now wasted and destroyed'.[8]

The Leighe continued across the high road from Mere to Shaftesbury at the present Two Counties Farm, and led into a tract identified in 1568 as Haselholt, or the wood of hazels. The bounds now turned south-eastwards past *Pimperleghe*, a name still found as Pimperleaze, another name which may be associated with trees. The 1568 bounds note here the Queen's Oak, 'standing by the marsh near to Haselholt pound' and the south end of Barrow Street Lane.

Another prominent landmark further on was the *Hore Apeldure*, otherwise the 'grey apple tree.' It will be noticed that many of these names indicate a dominantly wooded landscape. By 1568 this had become 'of old time called Hore Apeldore', and the name had been replaced by that of Gutch Pool, which places the old apple tree at the point where the River Lodden crosses the county boundary and leads down to the farm of this name. The name seems to indicate a channel or stream, with a pool giving the appearance of having been 'scooped out'.[9]

The old apple tree, or the river pool, were to be found in the middle of a stretch of the boundary which separated the forest from the woods of Mere. Later in the century the boundary would be modified by the laying out of a deer park on the Mere side by Richard, Earl of Cornwall, who in 1253 had built his castle overlooking the town. The park, which enclosed an area of some 1200 acres, was bounded on the Gillingham side by a double hedge with a ditch and mound, a prominent feature some 22 yards wide.[10] To the south of the ditch was *Kurhigge* or Cowridge, 'the ridge where cows are pastured' a hillier, and perhaps more open part of the forest rising to 400 ft. The 1568 account fills in some further names along the next stretch: *Newgate Bushes*, the north-east end of *Nettyates Lane* (Bridewell Lane), and *Wethers*, or Withies. At this point the 1225 survey gives us the earliest recorded mention of the name of Motcombe as it describes the boundary between the forest and the Wiltshire manor of Sedgehill.

By this point the boundary was climbing again to the higher ground of Knapp Hill, to descend to *Frengore* or Ferngore, a name meaning an oddly shaped piece of land with ferns, perhaps a reference to the kinks in the course of the boundary at this point. The boundary was crossed here by the Motcombe to Semley road. Close by on the other side of this road is *Soulescombe*, a name suggesting a valley or gully.[11] From Soulescombe the boundary turned southwards as it climbed to the heights of Kingsettle. This always seems to have been a heavily wooded part of the forest, the edge of the Greensand forming the boundary between the forest and the Semley manor of the abbey of Wilton. For some miles to come, the bounds were to be entirely those of monastic lands. The derivation of the name 'Kingsettle' is of some interest, one explanation being that it refers to the high, lofty position of the landmark, at 800 ft with wide views across the forest. Other explanations link the name to the royal ownership of the forest, suggesting some sort of royal lodge, perhaps similar to King's Court in Gillingham.[12]

Around Shaftesbury and Duncliffe

From Kingsettle the boundary followed the higher ground southwards, the 1568 account noting a *Pyle Cross* and the lower part of Little Down. By this point the boundary was dividing the forest from the lands of another abbey, that of St. Edward of Shaftesbury. Below Little Down was *Wearmwell* or Wormwell, known in recent times by the name of Cowards Shute.

> From Wearmewell by the boundaries of the counties as far as Vroggemeare along the valley. From Vroggemeare as far as Koggemanestone along the valley. From Koggemanstone always along the valley as far as Radewell.

This brought the bounds to the foot of the hill on which the town of Shaftesbury stands, although the 1225 account makes no mention of this flourishing abbey town divided from the forest by the slopes of Castle Hill. The 1568 survey tells us that 'the bounds lead to a cross standing near to the highway of Towthill at the town's end of Shaston.' The three accounts of 1225, 1300, and 1568 then all specify three points close to each other dividing the town from the forest hamlet of Enmore Green. These are *Vroggemeare* or Frogmere, a name which suggests a pool or a pit, and by 1568 had become a 'pool of old time'. Close by was *Koggemanstone* (or Cockemanston in 1300 and 1568), a boundary mark perhaps associated with the Cokeman family; and *Radwelle* or Radwell, a name also relating to a well or pool.[13]

From Shaston or Shaftesbury the forest bounds led westwards, once more into open country. More monastic property lay to the south, belonging to the Warwickshire abbey of Alcester and the Normandy abbey of Bec. By 1568 this had become known as the Liberty of Alcester. The 1299 and 1300 perambulations mention the Bitene, a valley bottom, and soon after the bounds start to climb to the peak of *Dunkweie*, or Duncliffe Hill, a 'dark hill or escarpment'.

Here the bounds of the forest met those of the manors of Stour Provost and East Stour (Plate 5). The former was a manor of the Norman abbey of Preaux, which founded a cell at Stour Provost.[14] East Stour at this time belonged to Shaftesbury Abbey. After Duncliffe the forest boundary descended northwards, following a stream later known as the Blind Lake, dividing the forest from the wood of William Cusin, who held the manor of East Stour from Shaftesbury Abbey. There is no mention in 1225 of a king's highway between Shaftesbury and Sherborne, the later Sherborne Causeway, but the crossing point at Black Ven (the 'dark fen') is given. From here the boundary followed the Sete, the stream which leads into the Lodden.

From this point the 1225 account differs from the later ones. In the 1300 and all later accounts, the boundary follows the Sete into the Lodden and so into the Stour, turning upstream towards the town. But the 1225 record is clear that the boundary leaves the Sete and instead follows a ditch on to the high road. This can only be the road from Shaftesbury. The outcome is to leave the land of the manor of Ham, by that time in the hands of Montacute Priory, outside the forest boundary.[15] The last part of the account tells us:

> And from that road as far as the stone bridge. From the stone bridge as far as Kings Bridge. From Kingsbridge up the stream as far as Huntingford.

From this point the survey is brief, perhaps because the boundary was now entirely within the manor of Gillingham and there was less need to define it in detail. The stone bridge must be Lodden bridge, its robust structure perhaps reflecting the amount of traffic passing between the town and King's Court. King's Bridge therefore becomes the town bridge, recorded in 1300 and all later records as Barnabie's bridge. The boundary now follows the Shreen upstream. There is surprisingly no mention of further fords or bridges, but they occur in the later accounts. Bay bridge is the *Pulverford* of 1300 and the Lodbourn bridge of 1568, 'of old time called Pomweyford'. The 1300 survey refers to the river as the Mere Water, hence the bridge further upstream at Colesbrook becomes the bridge at Mereford. By 1568 this too had changed its name, becoming the Powridge bridge, 'which of old time was called Mereford'. The 1568 account mentions the boundary passing Bengervill or Benjafield as the boundary returns to its starting point.[16]

CHAPTER 3

THE ROYAL VISITS

On March 13 1207 King John began a tour which took him from London via Farnham to his hunting lodge at Fremantle or Fremanton, near Kingsclere, and thence on to Winchester, where he arrived on March 20. The next part of his journey took him to Clarendon, Cranborne, Bere Regis, Powerstock, and so on to Exeter. From here he turned eastwards, journeying through Sherborne to reach Gillingham for an overnight stop on April 4. He subsequently travelled on through Clarendon and Fremantle once more to reach Windsor on April 13 (Figure 2). His Gillingham

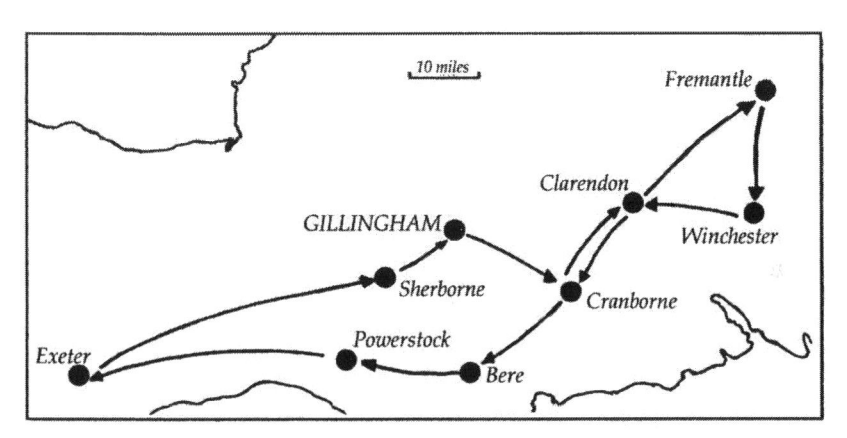

2 Part of John's itinerary for 1207, showing places stayed at in and around Dorset on the outward and return journeys.

stop was one of four visits made to Gillingham that year. Earlier in the year in January he had stayed at Gillingham on his way from Dorchester to Devizes, and a few days later returned to the town travelling from Sturminster Newton to Marlborough. He appeared at Gillingham yet again in July en route from Charterhouse to Clarendon. Between 1204 and 1214 he visited Gillingham on at least 25 occasions. The dates of John's known visits are:

1204	June 23-26; November 18; November 30; December 3-6
1205	November 10-14
1206	January 12-16
1207	January 25; February 6; April 4-5; July 24
1208	March 1; September 28-30; November 14
1209	July 3-4; September 23; December 15
1210	October 3; October 7
1211	February 12; February 15
1212	October 16
1213	March 15; July 8-9; August 1
1214	December 6-8

The destination of these visits was the royal hunting lodge or palace which stood outside the town of Gillingham on the edge of the forest, and known to later generations as King's Court. Across this decade, there was no year when Gillingham was not visited at least once.[1] Gillingham was one of many places visited from time to time by the Angevin kings. This was a period when the entire court was constantly on the move, its journeys becoming a series of 'progresses' linking together the many royal castles, palaces, and hunting lodges. The court had yet to settle into a permanent home at London or Westminster, the principle being that 'the court is where the king is.' One later antiquarian wrote:

> The King had many Palaces within the Kingdom, and used to hold his Court and celebrate the high Festivals of the year at one or other of them, as he thought fit.[2]

Constant travel around the country enabled the king to foster loyalties amongst his barons, hand out justice, hear petitions, raise taxes, enforce the forest laws, and reward deserving followers. The continuing business of government whilst on the road included the signing of letters, and the issuing of writs and instructions. Courts of eyre and assize had to be attended and justice dispensed. New statutes or laws might be made whilst on tour, for example the famous Assize of Clarendon, enacted by Henry II from his hunting lodge in Wiltshire. The royal party often sought to stay at castles or lodges close to forests and deer parks, where 'they put from them the anxious turmoil of the court, and take a little breath in the free air of nature.'[3]

Little is known in detail of the presence of the Normans or earlier Angevins in Gillingham. There is a tradition of William II meeting with Archbishop Anselm at Gillingham in 1094. Henry I signed a

The Royal Visits

charter granting the manor of Biggleswade to Lincoln Cathedral whilst at Gillingham in 1132, and there is a reference in Stephen's reign to the signing of a charter concerning the burgesses of Taunton.[4] John spent much of the first few years of his reign in France, and only began regular tours of England in 1204. He set a punishing pace of travel, his retinue averaging around twelve miles a day over his reign, with stops of more than three or four nights being uncommon. In Dorset, the castle at Corfe with its 60 recorded days, was the most popular venue, followed by Gillingham, with 48 (Figure 6, p.31). While Gillingham was popular as a Dorset venue, it was by no means frequented as much as Fremantle, Marlborough, or Clarendon, which acted as staging points for the more extended itineraries. John's first visit to Gillingham in June of 1204 involved a stay of four nights and was marked with considerable flourish:

> for the expenses of the king hunting at Gillingham, and for a feast for the poor at the king's first entrance into Gillingham, and for necessary expenses made by Ralph the park keeper, and his associates with the king's hawks, horses, and his pages, £7 3s 6d.[5]

John's son, Henry III, journeyed less often than his father and stayed at places longer. His venues were more likely to include holy sites or places of pilgrimage.[6] Gillingham does not appear to have been a favourite place, even though he authorised considerable expenditure on the residence. In 1220, and again in 1222 and 1223, during his period of minority, his court passed through Shaftesbury, but there is no mention of Gillingham. In June 1236 the court stayed at Corfe, Sherborne, and Cranborne, but Gillingham was passed by. There are several instances of wine being ordered for the residence at Gillingham, but some of it may have been intended as gifts rather than for royal consumption.

In July 1250 letters were signed at Gillingham, indicating the king's presence. In 1252 letters were signed at Gillingham between December 9 and 12. In December 1256 the king's huntsman was required to take deer from the forest of Gillingham for the approaching feast of Christmas, but the king actually spent the festival at Guildford on his way back to Westminster from Wiltshire. Edward I signed letters at Gillingham during visits in 1278, 1289, and 1297. Thereafter there is no record of royal visits to Gillingham. During his reign and after, Clarendon was the principal west country place for royal sojourns, and comparatively few trips were made into Dorset or Somerset.[7]

A king on the move

The royal retinue which accompanied John on his visits consisted of the entire court and much of the machinery of government. Besides the king and his family there were the officers of state from the different departments such as the Exchequer, Chancery, and Wardrobe. It has been estimated that even at the least busy times of John's itinerary, there could have been at least 500 individuals in the royal retinues. The household knights may have numbered up to 200. Besides the royal officers, knights, huntsmen, and active members of the progress, there were also the many wives, families, and servants. The touring retinue of William Marshal, Earl of Pembroke and principal counsellor to John, may alone have numbered 40 to 50.

Protection on the road came from the accompanying household knights with their pages and retainers. Lords and magnates close to the king, or seeking his attention for business or favours, would join the royal party with their own retinues. Further down the scale of society would be suitors and petitioners seeking royal justice, and paupers seeking alms from the king's table. The domestic retinue included the royal cooks, stablemen, and blacksmiths. The many horses in the progress would have included palfreys (riding horses), sumpter or baggage horses, and *destriers* or war horses. In April 1212 the Wardrobe alone moved around with a dozen carts, eighteen carthorses, and 77 palfreys.

A considerable part of the entourage was made up of the royal baggage, hauled from one venue to another. This had to include the king's personal possessions, the equipment of the royal kitchen, and all manner of artefacts including the royal bathtub. Much of the baggage party consisted of carts, trains of sumpter horses with their carters, and pack men. Being slower than the king's party on horseback, the baggage trains might take a more direct route or set off earlier, or wait for the king to return from a hastily arranged detour. In April 1212 we hear of payments to six sumpter men, ten carters, and twelve stable boys in charge of 51 palfreys, who had been obliged to tarry at Gillingham while the king was elsewhere.[8]

A further, essential, component of the royal retinue, but often staying separate from it, was the hunting party. This was itself a gathering of considerable size, consisting of huntsmen, fewterers (dog handlers), and falconers, together with their many dogs and horses. The hunting group tended to shadow the king's journey and was often to be found a few miles from the king's location, or perhaps might arrive at the hunting venue ahead of him. Its size might vary according to the king's needs and the hunting season, but rarely had fewer than 150 dogs. In September of 1212, while the king was moving around the forests of the Midlands, the hunt had 300 greyhounds, 16 boarhounds, 9 *bercelets* (hounds hunting by scent), and 64 handlers. A few weeks later, at Bath on 17 October, it was rather smaller, with 114 greyhounds. It might be significant that the previous night (October 16), the king stayed at Gillingham, continuing on to Bath the following day, and he may therefore have seen something of this gathering even if he did not make use of its services for hunting.[9]

The size of the travelling court would have considerably exceeded the size of the population of a small place like Gillingham at this time. The arrival of the king and his entourage, or even just of the hunting or baggage parties, would have been exciting, colourful, and 'a spectacular and threatening moment in the life of any community.'[10]

The King's Court hunting lodge itself would have been barely large enough for the king and his immediate household and attendants. The greater part of the party would have had to pitch its tents and coloured pennants around the edge of the park and the clearings in the forest along the Lodden and the road to Shaftesbury. The town would have welcomed the merchants and traders in the retinue who might bring materials and fancy goods otherwise rarely seen in the town, and there would also be opportunities for the town's own craftsmen to market their products to the visitors. However, the overall presence would also have spread alarm, causing people to look to their crops, their livestock, and their womenfolk.

For the local population, the high point of any royal visit to King's Court lasting more than a night or two would have been the announcement of a royal feast, most likely to be held at Christmas or some other major festival. The Angevin kings were lavish in their hospitality, and feasts or banquets also gave the king the opportunity to meet and gain the loyalty of local barons, knights, or clerics. At these occasions, an invited guest might be fortunate enough to be promised gifts in the form of deer or timbers as a reward for favours or services. It was also expected that during such a stay the poor of the locality would be provided for, and so for many Gillingham people, the great moment of the visit would have been the opportunity to feed off the remains of the royal feasting.

A great feast

John's great feast for the poor of 1204, described above, may have been remembered for years to come. Alongside the many common people of Gillingham there would have been numerous distinguished personages, some forming part of the travelling court, others of more local origin. Although such a list is speculative, there can be reasonable certainty about some of the individuals in the great hall at Gillingham in June 1204. We might expect to have found John's second wife, Isabella, only sixteen years old; William Marshall, Earl of Pembroke, a loyal supporter of the Angevin kings, later to have a manor (Sturminster Marshall) in Dorset which bears his name; Brian de Lisle (or Insula), a royal commander; Hugh de Neville, the Chief Justice of the forests; and Peter des Roches, John's Lord Chamberlain, who the following year would become Bishop of Winchester.

3 A barbecue in a forest (from the Luttrell Psalter).

Among the more local guests we might expect to see would be William Longespee, John's half-brother and Earl of Salisbury, who held Salisbury Castle and whose tomb can be seen in Salisbury Cathedral; and William Montacute, the sheriff of Somerset and Dorset, the official responsible for making many of the arrangements for the king's visit. A guest who is almost certain to have been present was Marie, the abbess of Shaftesbury and John's aunt. Some years earlier John had granted her the right to take two cartloads of brushwood daily from Gillingham forest, and it would be surprising if other favours from the forest, such as deer or other beasts, were not readily allowed. Some authorities have identified Marie as Marie de France, the earliest known woman writer in French, who created verse

The Royal Visits

romances on chivalric and folktale themes, including versions of Aesop's fables. Since much of her work would have been written at the abbey, such a connection would give Shaftesbury and Gillingham a link with an early stage in the development of French literature.[11]

Some idea of the scale of provisions needed for such a feast can be seen in the requirements for a Christmas feast at Winchester in 1206, when the king ordered 20 oxen, 100 pigs, 100 sheep, 1500 chickens, and 5000 eggs.[12] Edward I stayed over Christmas at Gillingham in 1275, for which 15 tuns of wine were required, and 36 oaks were to be felled to make brushwood and charcoal for the duration of the stay.[13]

After the accession of Edward I in 1272, the royal visits were much fewer, the last known visit being in 1279. One reason was that by this time kings were making fewer journeys; the court was becoming less itinerant and not so inclined to visit outlying parts of the country as often. While the number of royal residences in the fourteenth century was much the same as in the previous century, the distribution changed. Outlying residences were abandoned in favour of newer locations closer to London, particularly in and around Windsor Forest, but also in the New Forest. More attention was given to those residences with the potential for expansion and enhanced standards of comfort. Clarendon became the most westerly of the residences and a favoured place of Edward II and Edward III, both of whom spent considerably on the buildings and park. By the fifteenth century there were altogether fewer royal houses in the country, and only one north of the Chilterns.[14]

The feasting and lavishness which had taken place during Angevin times would not have been considered unusual, and many would have regarded it as necessary. In order for a king to demonstrate power and authority, it was thought essential that there should be a visual display of wealth and ostentation, and monarchs would be judged on the scale of largesse they could provide. Even a court which was itinerant was expected to display some intermittent grandeur. Gillingham could play its part in this alongside the many other venues on the royal tour.

Overleaf, pp.24-25:
4 The forest of Gillingham in the later Middle Ages, with King's Court and the deer park. This map shows many of the places mentioned in this book.

THE FOREST OF GILLINGHAM
IN THE LATER MIDDLE AGES
with King's Court and the Deer Park

- Enclosed ground and assarts
- Wooded areas
- Settlements and dwellings

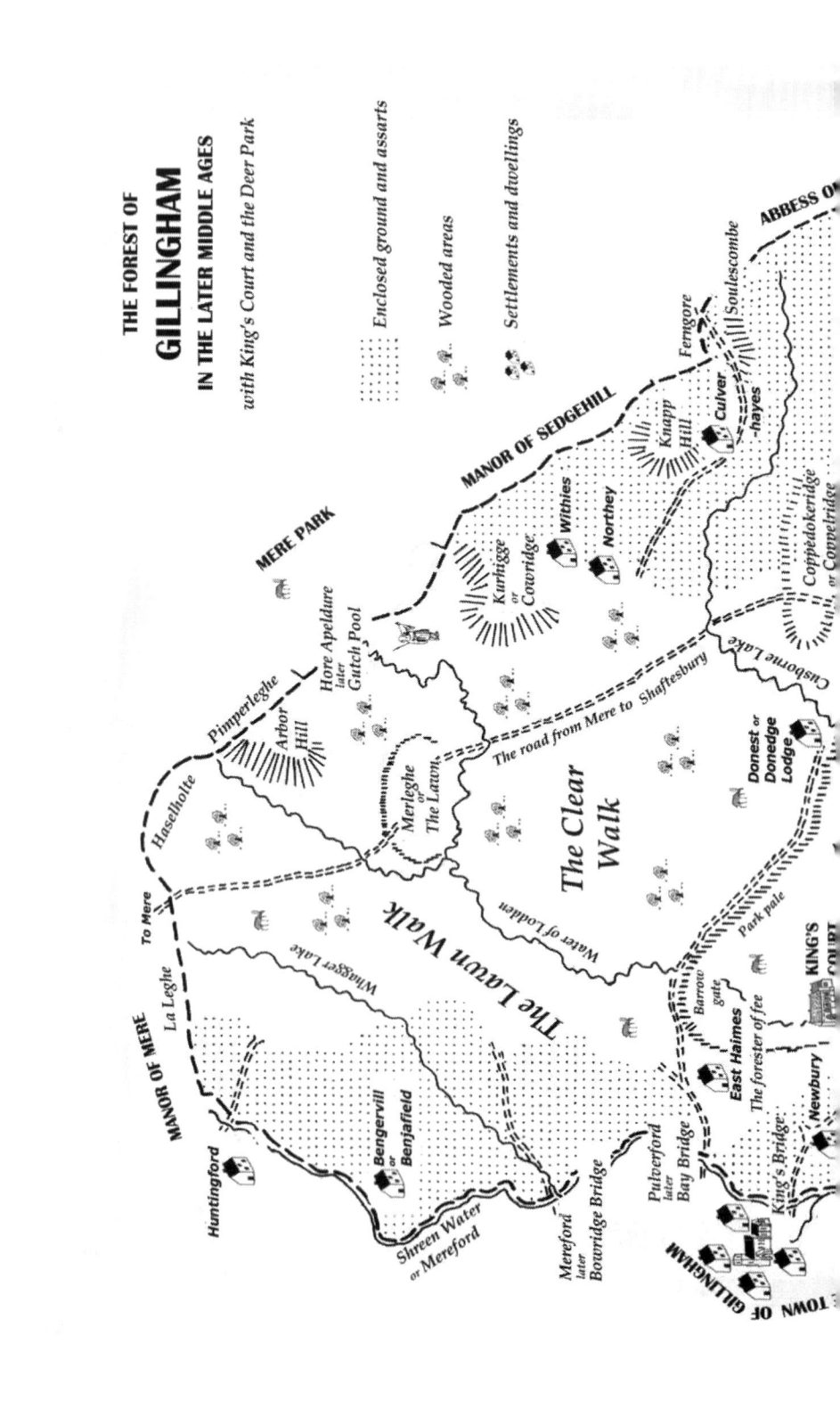

MANOR OF SEDGEHILL

MERE PARK

ABBESS OF [...]

Ferngore
Soulescombe
Knapp Hill
Culver
Withies
Northey
Kurhigge or Cowridge
Coppedokeridge or Compelridge
Hore Apeldure later Gutch Pool
Pimperleghe
Arbor Hill
Merleghe or The Laun
The road from Mere to Shaftesbury
Cusborne Lake
Donest or Donedge Lodge
Haselholte
The Clear Walk
To Mere
Water of Lodden
Park pale
MANOR OF MERE
La Leghe
Whagger Lake
KING'S COURT
The Laun Walk
Barrow gate
Huntingford
Bengervill or Benjafield
East Haimes
The forester of fee
Newbury
Shreen Water or Mereford
Mereford later Bowridge Bridge
Pulverford later Bay Bridge
King's Bridge
TOWN OF GILLINGHAM

Map of Shaftesbury and Surrounding Area

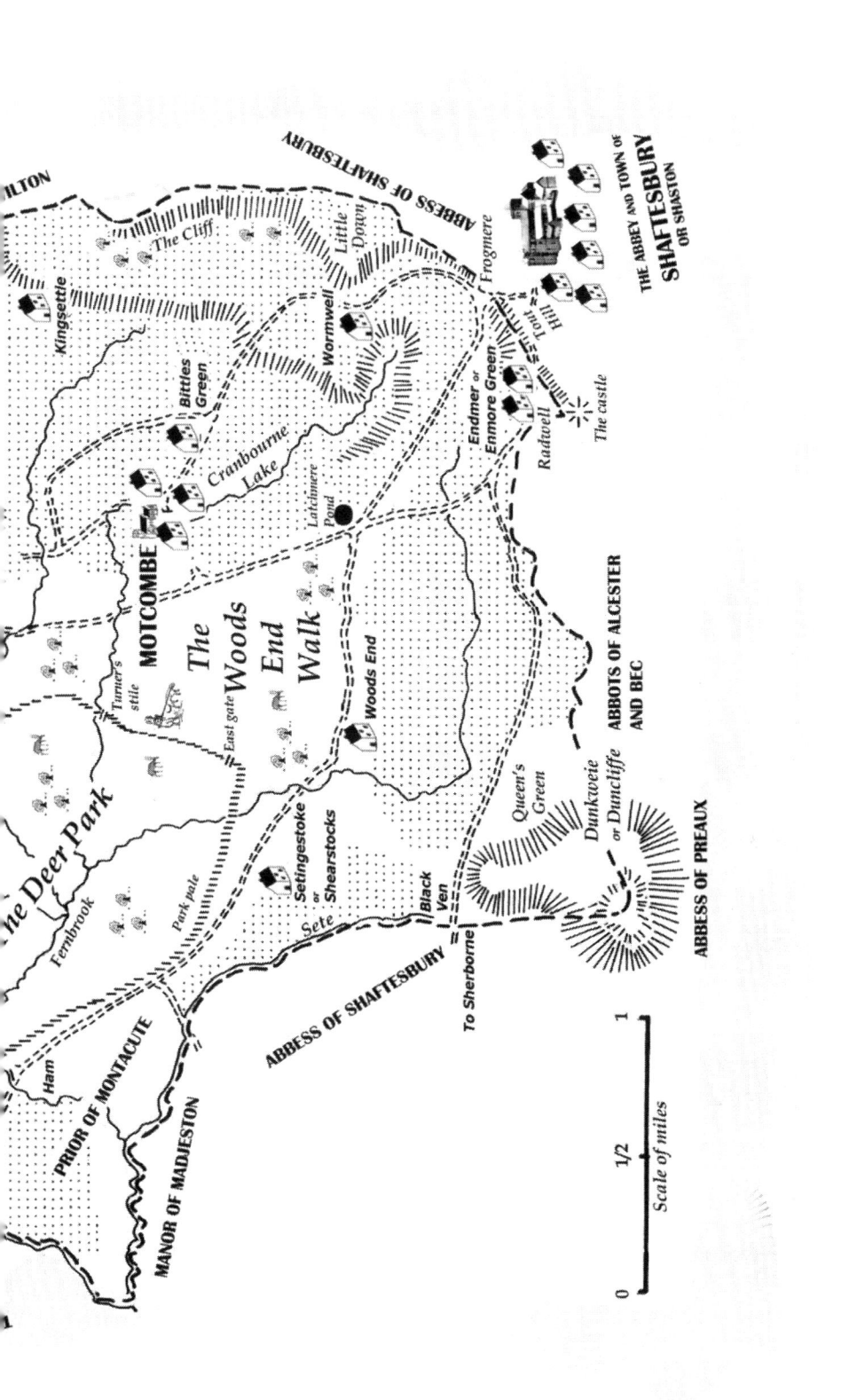

THE ABBEY AND TOWN OF SHAFTESBURY OR SHASTON

ABBESS OF SHAFTESBURY

ABBOTS OF ALCESTER AND BEC

ABBESS OF PREAUX

ABBESS OF SHAFTESBURY

PRIOR OF MONTACUTE

MANOR OF MADJESTON

Locations shown:
- HILTON
- The Cliff
- Kingsettle
- Little Down
- Frogmere
- Bittles Green
- Wormwell
- Cranbourne Lake
- Endmer or Enmore Green
- Tout Hill
- Radwell
- The castle
- MOTCOMBE
- Latchmere Pond
- Turner's stile
- East gate
- The Woods End Walk
- Woods End
- The Deer Park
- Fernbrook
- Park pale
- Setingestoke or Shearstocks
- Ham
- Sete
- Black Ven
- To Sherborne
- Queen's Green
- Dunkweie or Duncliffe

Scale of miles: 0 — 1/2 — 1

CHAPTER 4

KING'S COURT: PALACE AND PARK

Once past the ditch and bank of the deer park, a royal party would cross the bridge over the Fernbrook later known as King's Court bridge, and would be greeted by the only building of any substance within the forest. This was the royal 'palace' or lodge, built entirely for the accommodation and pleasure of the king and his party. In John's time the building was usually referred to as the 'king's house', the name 'king's court' being first recorded in 1253.

The earliest illustration of the king's court is its depiction on the forest map of 1624. By this time the buildings had long since disappeared, and the site is shown as a square-shaped earthwork marked as 'kinge's courte'. The Dorset antiquarian, John Hutchins, in his *History of Dorset* gives this description from a century and a half later:

> In the forest was anciently a palace, built by the Norman or Saxon kings for their residence when they came here to hunt. It stood half a mile east from the church, on the way from Gillingham to Shaftesbury, near two small rivers, on a level ground, encompassed by a moat, now dry, in some places 9 feet deep and 20 broad.[1]

Hutchins may be the earliest writer on Gillingham to refer to the house as a 'palace'. He goes on to give some further specific dimensions:

> There are traces of a rampart that appears to have been 30 feet thick, but it is now little higher than the area it encloses. This area, in which the house stood, is 320 feet long and 240 broad. The foundations are still to be seen, though not a stone of it is left. It was built in one corner of the area, about 20 feet from the rampart, in the form of a letter 'L'; the length of the body of the letter is 186 feet by 80, the foot of the letter is 48 feet by 40. The area of the house contained 16,800 square feet, and the whole enclosure is 3 roods and a half, or three quarters and half a quarter of an acre.

There is now no way of verifying the accuracy of some of Hutchins' dimensions, but their specific nature suggests that they were based on detailed measurement from observations possible at the time. Hutchins obtained a lot of his information on Gillingham from the vicar, William Newton. Since Newton died in 1744, it is likely that this information came to Hutchins some decades before his *History* was published.[2] The measurements given are those of the palace at the end of several decades of development; the layout of the palace in earlier times may well have been different.

A more recent survey was that carried out by the Royal Commission on Historical Monuments for their 1972 volume on North Dorset (Figure 5). This gives a rectangular area of 300 by 170 ft, and a bank of 50 ft width and 4 ft height, with scarps and platforms on the south and south-eastern sides. The ditch is described as up to 60 ft wide and 5 ft deep, and the outer bank 3 ft in height. The survey places an entrance at the south-west corner, with rectangular platforms suggesting former gate towers.[3]

While there is some difference between the Hutchins and RCHM dimensions, both accounts agree on the basic scale and layout of the feature. The RCHM survey depicts an entrance in the south-west corner with gatehouse, while Hutchins suggest an L-shaped cluster of buildings. The most likely location for the buildings would be the north-east corner facing towards the gatehouse, and some confirmation of this could be the remains of platform-like features in these areas.

The site has never been excavated, and it is likely that without archaeological investigation the layout of the site will always remain speculative. However, a royal hunting lodge of similar scale and layout was excavated at Writtle, Essex, in 1955. This was built by John in 1211 and resembles Gillingham in its rectilinear layout with moat, gatehouse, and hall. There was additionally a 'vivary' or fishpond beyond the moat, fed from a nearby stream.

King's Court: A palace fit for a king

The royal residences of Angevin times were invariably based around the 'great hall', used as a general gathering point for the king, his household knights, visitors, and guests. Here, banquets could be held,

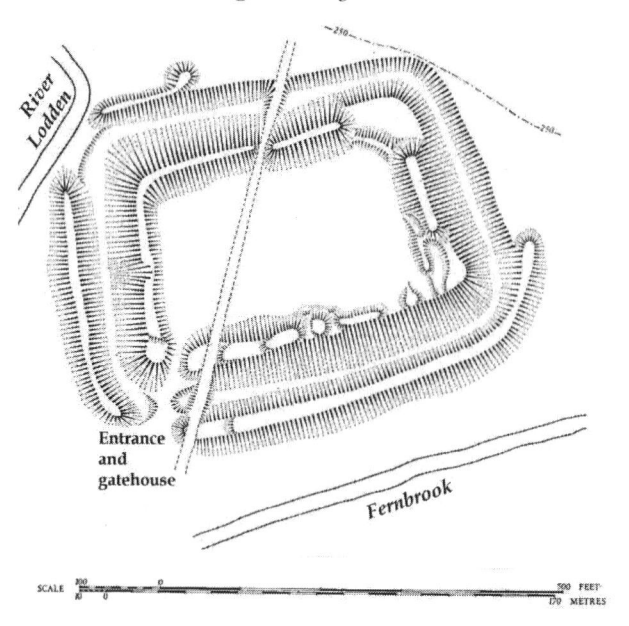

5 The King's Court earthworks (after RCHM)

the fruits of hunting or falconry set out for all to admire, and some of the servants might bed down. The most likely layout for this was that of the aisled hall, most likely to have been built of stone. Inside the hall, the roof structure would have been supported by great stone pillars. The interior would have been open to the roof, with a great hearth or fireplace in the centre, and smoke escaping through a hole in the roof. At one end a raised dais would have thrones for the king and queen and a long table. At the other end of the hall, doors would give access to a buttery and pantry. These in turn would be joined by a walkway or cloister to the detached kitchen. This was the layout followed at the larger royal palace at Clarendon in Wiltshire.[4]

While this is the most common type of medieval hall, it is possible that Gillingham may have been of a different type. The Angevin kings favoured structures with the great hall on the first floor, the ground floor being an undercroft for the storage of wine and other commodities. In this case the hall would have been reached from the outside by an external staircase. This layout was used at another of John's palaces at Cranborne. Here, the medieval hall still survives as part of the later manor house.[5]

The royal chambers were housed in separate buildings (Plate 1). We might expect the king's chamber to be not only his bedroom but also his office and workspace. Besides his bed and chairs, the room would contain benches and tables where the king could eat separately from the rest of the household and meet with his councillors. From here orders could be dictated to scribes, writs signed, taxes received, petitions could be considered, and visitors could be seen in private. The king also had his own chapel, and from the information given in building records, his Gillingham chapel would seem to have been on the ground floor with the chamber above. Even in royal and aristocratic houses complete privacy from the rest of the household was hard to obtain. An order of July 1250 (p.33) required a new chamber, suggesting the addition of a cross-wing to the main hall, giving it an L-plan, consistent with the shape of the site described in Hutchins.[6]

The queen had separate chambers, also with her own chapel. There was also a chamber for the king's son, and another for guests. The different buildings would have been linked by walkways or cloisters, with upper floors accessed by external as well as internal stairs. Outbuildings included stables and eventually a gatehouse with porter's lodge, and a lodging for the king's chaplain. There may have been separate gardens for king and queen, with private access from their respective chambers. These lesser buildings were of timber construction using the plentiful oaks of the forest, but there is evidence of stone being used in the rebuilding of the chambers. The kitchen would have been a detached building, perhaps square in shape, with a hipped roof.

The degree of domestic comfort to be found in the Gillingham palace cannot easily be judged by modern standards. This was a time when glass was scarce, even in royal houses, windows being closed by shutters held in place by wooden pins. However, the use of glass is recorded in the rebuilding of the king's chapel and chambers. Floors were normally covered with straw or rushes. Wall tapestries were uncommon, and at this period having too many might be considered over-indulgent. Henry III favoured brightly painted timbers and wainscotings for many of his houses, and these were part of his rebuilding programme for Gillingham. The king's favourite colour was green, which he often combined with gold stars or other shapes. Such green and gold backdrops were then further adorned with paintings of biblical histories or of knightly scenes.[7]

However, the level of interior finish at Gillingham may have been modest compared with that at Clarendon, one of Henry's major palaces. Here the most novel, fashionable, ideas in painting and sculpture abounded, with no expense considered too great. At Clarendon the windows to the hall and main buildings were of painted glass, the chambers were adorned with historical and allegorical scenes, and the ceramic tiled floors covering the royal chambers and chapels were viewed as a wonder of the age. Here his tastes included scenes of Alexander the Great, and depictions of the great duel between his uncle Richard I and Saladin.[8]

An earlier writer on the palace at Gillingham has given this evocative description:

> Just a glance around Gillingham court before finally leaving it. Behind us is the gateway with porter's lodge over it; in front is the hall, the kitchens and their offices are out of sight, but the chimney shows above the chambers on that side. The chambers are detached, two storeys in height, some of the older houses of timber framed with plaster panels rich in colour, and roofed with shingles; the king's, queen's, and the newer chambers of stone. Then there are the corridors with the external staircases. The whole group has a quaint and picturesque appearance, the picturesqueness being heightened by the vines trained over the face of some of them, the bright green of the foliage contrasting with the coloured ornament with which the outside of some of the buildings is adorned.[9]

King's Court: the king's works

The 'palace' as described above was not the result of a single building project, but evolved through several stages of construction. Gillingham was one of numerous royal residences on which John and Henry III lavished large amounts of money (Figure 6). John had over twenty royal houses spread around the country, and his spending on these residences was second only to his spending on castles.[10] Gillingham is one of a handful of residences referred to historically as 'King John's house'. Not all of these are associated with John, but Gillingham is one which is entitled to be so called.

The development of King's Court can be followed through the various references in the royal records to building works and alterations. The residence must have existed in some form in the mid-twelfth century since in the Pipe Rolls we learn of regular payments of 1d a day (30s 5d per year) to the keeper of the 'king's houses' in Gillingham.[11] The use of the term 'houses' suggests a structure which had several elements or was evolving piecemeal, but its size or appearance at this time is not known. It underwent extensive rebuilding in the early years of John's reign, since in 1199 three lime

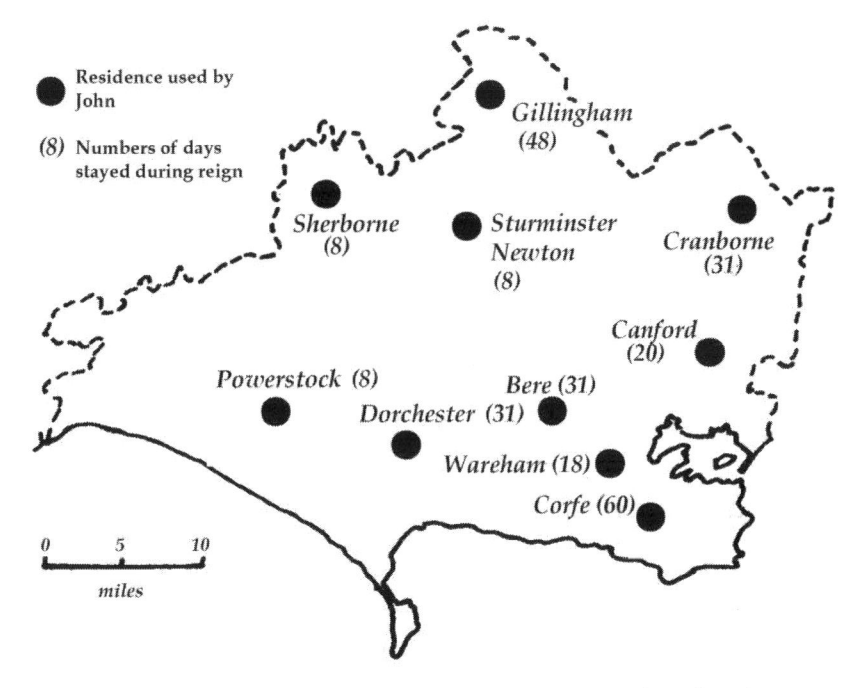

6 The royal residences of Dorset. The figures indicate the total number of nights stayed at each residence by King John.

kilns were ordered, presumably to make lime from which mortar for £129 7s 4d is recorded. The Pipe Rolls also record the names of the persons called upon to inspect the building works. All this amounted to the major rebuilding which was the occasion for the triumphal entry of the king into Gillingham in 1204 and the great feast already described.[12]

The improvements and alterations to royal houses seem to have been very much the decision of the king, who instructed the warden or sheriff of his needs, the costs being paid out of the farm (income) of the county. Once Henry III took over his own affairs from his regents, spending on royal residences increased further. Over his reign he spent £28,000 on his houses, about twice the rate of expenditure of John. In 1228 Hugh de Neville was instructed to spend £8 on repairing the king's house at Gillingham, Feckenham (Worcestershire), and Poorstock (Powerstock), and in 1233 there was an order to roof and repair the Gillingham houses. In 1237 Robert of Aundeley, the forest warden and manor steward, was again directed to repair the houses, which had been damaged by tempest. In September 1239 he was instructed to roof the kitchen, repair the gutters, and provide a 'good hedge' around the king's court. The latter may have formed an inner fence within the main enclosure, to enhance the privacy of the royal household. In May 1241 67s 10d was spent on repairs to the king's buildings.[13]

There was some development of the chapel facilities at this time. In 1239 a suitable chaplain was to be engaged to minister daily in the king's chapel at a rate of 50s a year. Cracks in the chapel wall were to be filled and whitewashed, and the chapel door repaired. In 1241 the chaplain was allowed 2s for 4lb of wax for candles for the chapel. In 1244 the sheriff of Dorset was directed to buy vestments, altar cloths, and a chalice for the chapel. These are the earliest references to a resident cleric in Gillingham; until that time ministry in the town and at the palace would most probably have been provided by Shaftesbury Abbey.[14]

In 1244 Aundeley was directed to make a 'privy wardrobe for the king's use between the king's and the queen's chamber'. Until this time the chambers had been separate buildings, but this instruction suggests some type of structure to link them together. Such a room would have been a private room, dressing room, or room used for storage of personal materials or valuables.[15]

These alterations might seem surprising, considering that they came from a king who, from what is known, had not yet set foot in Gillingham. From 1249 even more ambitious ideas began to appear, clearly with the aim of producing a royal residence of some status. In October 1249 an order was made to rebuild the chapel, which was threatening to fall down, despite the repairs of just a few years earlier. The following January the foresters were to provide 40 oak trees for the building of what must have been a new chapel. Then in July of 1250 there came instructions for a much more thorough rebuilding programme, written during a two-day period of the king's visit. The work on the new chapel, which was now above the king's chamber, was to be completed, including six windows 'with columns in the middle'. The king's chamber was to have a fireplace and a further window, also divided by a column.

The references to windows and columns indicate a stone structure, no doubt with windows in the Gothic style, which was very much in favour with the king. Such windows may have had iron bars, following an attempt on the king's life at Woodstock when an intruder climbed in through an unbarred window. The queen's chamber was to be lengthened at the gable end, and she was also to have a chapel and a fireplace.[16]

Further direction from Henry's July 1250 visit involved considerable enlargement to the hall. This is described as 'at the head of our hall there, towards the east a chamber forty feet long and twenty-two feet wide, transversely towards the north with a chimney (fireplace) and privy chamber.' These dimensions would suggest a considerable extension built as a cross-wing to the existing structure. In the same month the forest steward was to provide a further 50 oaks for the king's carpenter. In March of 1251 80 logs were to be set aside for the king's works at Gillingham, 40 to be taken from the forest, and 40 from the park. In November a further 10 oaks from the park and 10 from outside the park were required.[17]

A costly enterprise

By this time, the cost of these works may have been taking its toll on the royal purse. In May of 1251 the sheriff was directed to take 50 marks (£33) from the Exchequer to spend on the works at Corfe and Gillingham. The following month he was to allocate 40 marks (£27) from the farm or income of the counties of Dorset and Somerset, and in September a further £60 was to be set aside. In November of 1251 Elias of Rabyn was directed to take good and diligent care for the completion of the king's works at Gillingham, Sherborne, and Corfe, and was to send for 100 marks which he was to receive from the sheriff of Nottingham; however, a further note indicates that this latter amount was directed to other works on the royal palace at Fremantle in Hampshire. An entry in the Pipe Rolls for 1251 indicates a cost of £198 3s 0½d for the various items of building work for that year.[18]

The work continued to move on throughout the rest of 1251 and 1252. In November of 1251 Ralph of Godmanneston, Robert of Wyke, Robert of Bertum (probably Bourton), and Richard Cressebcn were instructed to view the king's works at Gillingham. In January of 1252 the sheriff was to view the faults in the works at Sherborne, Corfe, and Gillingham, and make them good. In November a further 20 oaks were needed. In March of 1253 'in consequence of the removal of the bailiff of Gillingham and the delay of the king's works there', the abbot of Pershore, who was the king's escheator south of the Trent, became involved in making sure that funds were available to keep the work going. The following month it was directed that all the revenues from the manors of Gillingham, Bridport, and Somerton, except the tallage (tax) should be laid out on the king's works in Gillingham, suggesting that desperate solutions were needed to meet the rising costs. In June the completion of the work was entrusted to William of Monte Sorello.[19] In December of 1252, over his three-day stay, the king would have been able to admire the progress of his new works.

In December of 1253 major changes were ordered to the outside of the palace. Until this time the boundary would have been marked by a simple palisade, but now the entire boundary was to be ditched, ie moated, with a stone and mortar wall built to a man's height. The moat would have been regarded primarily as a decorative feature, one intended to enhance the appearance of the palace and provide a sense of completeness to the whole palace complex. The ditch was to have a bridge leading to a new gateway, which would have accommodation for the porter over its top. The 'great chamber', possibly the extension recently built to the hall, was to have a privy chamber and chimney, and was to be whitewashed. The hall was to be roofed, or maybe re-roofed. An almonry house was to be built, to also contain a chamber for the chaplain. Detailed instructions were given to complete the ornamentation of the chapels. In the king's chapel:

> Likewise to wainscote and illuminate or paint the King's chapel and chamber and to put on every side in the king's chapel and cause to be painted on the glass windows three images, to wit of the Blessed Mary, St. Edward the King and Confessor, and St. Eustace, and to make benches and forms in the said chapel.[20]

While in the queen's chapel:

> To complete the Queen's chapel with the chapel in honour of St. Edward King and Martyr, and St. Edward, King and Confessor, and to wainscote and illuminate the same chapel, and likewise the Queen's chamber, under which ... to make a new wardrobe for the Queen's use, with a chimney and privy chamber.

The decoration was also to apply to the chamber of Prince Edward:

> To whitewash and illumine the chamber of Edward, the King's son, and to make doors and windows to the same where necessary.

King's Court: Palace and Park

The references to stained glass windows demonstrate the great lengths and expense to which the king was prepared to go to enhance the splendour of a residence which he might use for only a few days each year. By his later years Henry III was becoming a less impetuous and unpredictable personality, more given to matters of piety and devotion, with a strong attachment to Edward the Confessor. The dedication to St. Edward the Martyr in the queen's window may have reflected a regard for the great saint of Shaftesbury Abbey, just three miles distant.

By this time the palace had reached its final form. One reason was that the king had now overreached himself, and was quite unable to pay for the works needed. In the absence of the king in May of 1255, the abbot of Pershore ordered the wages of the masons and workmen to be paid from the sale of quantities of grain held at Gillingham, Melksham, and Somerton. In November of 1256 the sheriff was to pay 30 gold marks to the king's workmen as part of the indebtedness owed by the king. By now it is likely that some of the alterations were being long delayed. In 1261 the sheriff was ordered to wainscot over the altars of the two chapels, and to 'make a certain bench between the king's hall and the kitchen to arrange the king's dinner on.' An expense of £4 18s is given in the Pipe Rolls.[21] As far as can be known, the king never visited Gillingham to see the impact of these later works to the palace.

The attention and expense lavished by John and his son on Gillingham over two generations had been considerable, and might be taken to indicate that this was a residence to which they attached special regard. However, it must be remembered that Gillingham was one of a large number of royal residences of Angevin times, all of which attracted high levels of royal spending. Around the time John was spending large sums on Gillingham, he spent £67 on his royal lodge at Cranborne and £1250 on Corfe Castle. Henry III lavished even greater sums on his palaces and residences. His £28,000 spent on all royal houses can be compared with £10,000 on the Tower of London, £15,000 on Windsor castle, and £40,000 on Westminster Abbey. At one point in his reign his debts were three times his income.

Nothing is known about how King's Court was used and occupied apart from the brief moments of royal presence. An interesting reflection is that, whilst its splendour was plainly visible to the royal and knightly elites who visited it, little of King's Court would have been known to the common people of Gillingham. Concealed behind the high park fence, with only the higher parts of its buildings projecting, the palace would have been largely unknown to many townspeople. Occasionally a glimpse may have been obtained when the gates were open, and from time to time the townspeople may have been invited in to enjoy the remains of a royal meal. But for most of the time King's Court would have been the great building behind the wall, known about, but known to few.

The deer park

To reach the palace, the party would have followed the boundary along its southern edge, before turning right over a bridge and passing through a gate in the bank at what is now the end of King's Court Road. The date when the park was first laid out is not recorded. We first hear of a park keeper at Gillingham in 1204, and in 1228 there is specific mention of a park and deer leap. It is likely that by these dates the park had already been in existence for some decades.[22]

Since hunting was the prime recreation of all within the knightly and higher orders of society, parks were regarded as a necessary symbol of lordly wealth and status. It has been estimated that there may have been around two, or even three thousand deer parks in England by the thirteenth century, but not all may have existed at any one time. Deer parks could provide sport for the lord and his followers, and also contribute to the lord's larder. Wherever emparkment took place, it emphasised the culture of lordly control of the countryside to the exclusion of others.

In Dorset some 60 deer parks are known. Many of these deer parks were in the far western and far eastern parts of the county, there being relatively few across the south. The nearest parks to Gillingham were those at Kington Magna, Stalbridge, and Tarrant Gunville.[23] In Somerset there were parks at Charlton Musgrove and Stoke Trister, while in Wiltshire deer parks could be found at Stourton, Zeals, Donhead, East Knoyle, and at Mere. The Mere park adjoined the forest of Gillingham along its northern boundary.[24] Gillingham was one of only four royal parks in Dorset.

While a forest could provide an appropriate environment for deer to roam wildly, a park provided a more enclosed area where they could be allowed to breed, safe from trespassers and poachers. The royal deer parks needed woodland dense enough to harbour quarry, but also smaller areas of thicket and open spaces to enable a degree of pursuit and capture. Other requirements included adequate drinking water from streams or ponds, and sheltered, secluded spaces where fawning could take place without interference. Some hunting might take place within the park itself, but the deer could also be released into the surrounding forest for a more protracted chase.

The nature of the park boundary, known as the *park pale*, was of some importance. As a deer can jump up to three metres vertically and six metres horizontally, the pale had to present the deer with a formidable barrier. This frequently took the form of a high hedged bank with a deep ditch, the bank being enhanced with stakes or pales to a height of three or four metres. This ensured that the deer park was separated from the lands beyond, access on foot or horse being only through clearly marked gates and stiles. On the inside, the steep-sided ditch would deny the deer the footing needed to leap the fence. An essential feature was the provision of gaps or 'deer leaps' in the boundary. These might take different forms, but were always designed to allow the deer free ingress while restricting their exit. One design was to build into the gap a sloping fence or ramp which could only be used by deer entering the park. In this way the stock of wild deer in the park could be gradually increased from the forest outside.[25]

The cost of maintaining the ditches, banks, palings, and gates could be a considerable outlay, but proper maintenance of the park was considered necessary as a demonstration of the lord's social position. The park keeper or parker might need to patrol the perimeter regularly or daily to ensure that gaps in the pale were promptly filled. The demand for stakes was such that, apart from providing browse or underwood for sheltering the deer, a park's timber resources might have been mostly used for the supply of fencing. At Gillingham it is notable that several of the royal orders for the taking of timbers from the forest require that they be taken from outside the park, not within it.[26]

The Gillingham deer park, once established, remained largely unaltered through the centuries. It is clearly shown on the 1624 map or 'plott' of the forest, and the alignments of the ditch and bank are well preserved in the pattern of present-day field boundaries. Some later names such as Lower Park, Park Farm, and Pale Mead reflect the existence of the former park. The ditch and bank enclosed an area of 760 acres, some four miles round, mostly east of the Lodden and occupying much of the countryside between Gillingham and

Motcombe. The size can be compared with that of smaller royal parks such as Ludgershall in Hampshire (c.250 acres), but also with the much larger parks of Guildford in Surrey (c.1,670 acres) and Clarendon in Wiltshire (c.4,300 acres). While much of the outline of the boundary is still traceable in later field boundaries, preservation of the actual bank and ditch itself is more limited. The best preserved section is a length of some 1½ miles on the eastern side of the park between Donedge Lodge Farm and Waterloo Farm (Plate 2). There were entrances to the park at Barrow Gate, at the King's Court palace to the south (West Gate), and at Turners Stile and an East Gate on the eastern boundary.[27]

7 A bowman waits for the deer to be driven towards him (from a French hunting treatise)

The substantial size of the park suggests that, besides its use as a facility for maintaining deer stocks, it was large enough for sport to take place within the park boundaries. While the park may not have been large enough for a protracted chase of the larger red deer, it was of a suitable size for more limited pursuits involving the smaller fallow deer. A popular, relaxed, practice of the time was 'bow and stable' hunting, in which the fallow deer could be driven towards suitably positioned nets. Here they could be finished off by dogs or by the bow and arrows of the carefully concealed lord or his lady (Figure 7). In his hunting treatise *The Master of Game,* Edward, Duke of York explains that

> if the hunting shall be in a park all men should remain at the park gate save the stable [ie station for bowmen] that ought to be set ere the king comes, and they should be set by the foresters or parkers.[28]

Deer parks were frequently stocked with a wider range of quarry than deer, including rabbits, hares, swans, pheasants, peafowl, and partridges. Rabbits were raised in *coneygarths* or enclosed warrens. Streams, ponds, and moats were well stocked with fish. All could be used to supply the lord's larder, or as gifts to reward his followers. Falconry was a frequent pastime in parks. There is no reason to suppose that the Gillingham park was any less diverse in its range of game or quarry than other parks of the time.

It might be significant that the hunting lodge or palace of King's Court was built within the perimeter of the park rather than outside it. This could suggest that the purpose of the park went beyond deer nurturing and sporting activities, and extended to leisure or decorative aspects, providing a pleasure ground for the king and his family. Such an interpretation of the park and its purpose is consistent with the lavish rebuilding of King's Court by John and Henry III, where the intention was to provide a royal residence in its fullest sense, including not just buildings but gardens, wooded aspects, and scenic views. The frontage of the palace would have been its northern and eastern facing sides, embracing views across the park and into the forest towards the higher ground of Kingsettle and the more distant downs beyond Mere. From the windows of the newly rebuilt royal chambers, the views might have taken in woodland cleared to provide gardens and ornamental wooded glades, with decorative fishponds fed from the Lodden or the Fernbrook. Such features would have been deliberately designed to enhance the appeal of the palace and its status as a royal residence.[29]

CHAPTER 5

FOREST GUARDIANS AND JUSTICE

On November 5 1222 Brian de Lisle (also known as Brian de Insula, meaning 'of the Isles'), was appointed warden or steward of the manors of Gillingham and Powerstock and of the king's houses there. De Lisle (d.1232) held the office for two years. The custody of the manor and houses at Gillingham always included the forest and park, which were treated as part of the manor. The appointment was at the king's pleasure, carried no land or property, and Brian de Lisle was to pay to the Exchequer a rent or 'farm'.[1] In return for this, and for looking after the king's interests in the manor and its forest, he would be entitled to any of the profits out of the appointment. These normally comprised the fines levied in the various courts, fees paid for pasturage or pannage, and the proceeds of any produce sold off the manor. These could be considerably in excess of the farm paid, so that stewardships or wardenships were offices which were highly sought.

Brian de Lisle was typical of the great barons, nobles, or captains in the royal favour who could readily secure offices such as forest wardenships. Gillingham was only one of numerous such offices which he held from time to time. He had been a mercenary captain under John, with a power base in Yorkshire, where he had commanded royal forces and several castles. He was described by the antiquary, William Dugdale, as 'one of John's evil counsellors'. Under Henry III he was again constable of several royal castles, and from 1222 to 1224 was Chief Justice of all the royal forests. In Dorset he held the manor of Blandford, and his name is perpetuated at nearby Bryanston.

It is likely that Brian de Lisle had very little to do with Gillingham in any direct way. He may have visited Gillingham as part of the retinue which accompanied the itinerant court, and is likely to have been at the great visit and feast of 1204. He took part in the 1224 perambulation of the forest, and during his time as Chief Justice he would have presided at the Dorset forest eyres, or county courts for the forest, where presentments from Gillingham forest would have been made. Like other nobles and high-ranking royal servants, he would have delegated the day-to-day running of the forest to local deputies and the foresters.

Justiciars and wardens

Even at the time of its greatest favour with John in the early thirteenth century, the royal retinue was present at Gillingham for only a few days in each year. For the rest of the time the forest had to be managed in such a way that the king's rights were safeguarded to the exclusion of all other claims or interests. The warden was directly answerable to the Chief Justice of the Forests, sometimes called the Justiciar of the Forest or the Chief Forester of England. This was one of the great offices of state of Angevin times. The Justices of the Forest sat on the commissions of forest eyre held periodically for each county. They were responsible for making sure that the forest revenues due to the king, for example from rents and fines, were collected and paid. They could investigate misconduct by wardens or other officials and seize their properties or bailiwicks.[2]

The office of warden was usually awarded to someone of high esteem in the king's sight, to a member of the royal family, or to a person of exalted rank as a reward for royal service. Like the office of Chief Justice, this was an office valued for its status and profits, with many of the duties being performed by lesser officials. These included oversight of the state of the deer and of the forest timbers; the apprehension of trespassers and encroachments; and presiding at local courts of attachment and forest eyres. The warden was responsible for royal instructions for the taking of timbers and their use in building work or as gifts to subjects. Tallies of expenditure had to be kept and submitted to the Household or Exchequer. As a reward for these services, the warden collected fees paid for herbage (cattle grazing) and pannage (pig grazing). He normally paid a fixed sum or farm to the Exchequer for the rights to collect these revenues, and kept any remaining income for himself.[3]

Another officer who frequently appears in the royal instructions for the management of forests is the sheriff. The sheriff was the chief administrative and judicial officer of the county, where he acted as the agent of the king in many matters. During the medieval period the shreivalties of Somerset and Dorset were combined as a single office.

In the absence of the warden, the sheriff had much of the oversight for the protection of the forests in the king's interests. During royal visits he would ensure that the king was comfortably provided for. He acted upon instructions regarding royal building works such as castles or hunting lodges, making sure that the work was properly completed and paid for. The sheriffs were closely involved in the rebuilding of King's Court.

The names of the wardens can be readily traced through the thirteenth century by their appearance in the Patent Rolls, Close Rolls, and other administrative records. Figure 8 is a list of the known wardens of Gillingham in the thirteenth century. Brian de Lisle was followed in 1224 by Hugh de Neville, who was Chief Justice and also had custody of the Dorset forest of Poorstock or Powerstock. Hugh de Neville was the son of Alan de Neville, who had been Chief Justice during the reign of Henry II. Hugh de Neville had held the office of Chief Justice many years previously during the reign of John, when he had set up his own version of the exchequer at Marlborough to deal with the large sums he raised from the royal forests in dues and fines. During his lifetime he held the office of sheriff several times. Along with Brian de Lisle he presided at the cutting back of the boundaries of Gillingham Forest to their earlier extent, as described in Chapter 1.

Other great barons or magnates who from time to time held the custody of Gillingham forest in the thirteenth century include John of Monemue or Monmouth, a baron of the Welsh borderlands and principal ally there of Henry III. Robert Walerand was a household knight and adviser to the king for much of his lifetime, and held many offices of constable, warden, sheriff, and justiciar. He was considered not so severe a justiciar as some of his predecessors. Roger of Clifford had his base at Kingsbury, near Tamworth, and in the 1260s was in command of the royal castles at Ludgershall and Marlborough. For a while he sided with Simon de Montfort against the king, but after reversing his allegiance he was rewarded with further lands and estates. He later joined a crusade under Prince Edward.

Henry III made great efforts to dislodge powerful magnates from offices of state and replace them with his household officials, especially French foreign favourites. In 1232 the forests of Gillingham, Clarendon, and Dean were committed to the Poitevin, Peter des Rivaux or Rivallis, who became Chief Justice of the Forest for life in the same year. He already held the offices of Keeper of the Wardrobe, the Chamber, and Treasurer of the royal household, as well as

1222	Brian de Lisle	1253	Ernaldus de Bosco
1224-29	Hugh de Neville	1255	Adam of Grindham
1231	John of Monemue (Monmouth)	1258	Robert Walerand
		1265	John de la Linde
1232-34	Peter des Rivaux (or Rivallis)	1267	Alan of Plugenet
		1268	Roger of Clifford
1236-47	Robert of Aundeley (or Dandely)	1269	Roger of Clifford / Alan of Plugenet
1247	Robert Passelewe	1271	Roger of Clifford / John de la Linde
1250	Hugh Passelewe / Ralph of Goddemanstone / John of Vernon	1273	Walter de la Linde
		1275	Guy of Taunton
1251	John of Vernon / Ralph of Goddemanstone	1276	Roger of Clifford / Guy of Taunton
1252	Adam Wymer / Robert Walerand	1293	Roger Lestrange

8 The wardens of Gillingham Forest in the thirteenth century.

numerous other stewardships.[4] Another Poitevin linked with Gillingham through the building works at King's Court was Elias of Rabayn, who was constable of Corfe and Sherborne castles at this time.

A warden who held the post for many more years than usual, and about whom little is known, was Robert of Aundeley or Aundely. Aundeley, first appointed in 1236, was reappointed in 1242 and held the office until 1247. He held no other similar offices, and his family name is traceable only in Oxfordshire and Northamptonshire, where a Maurice d'Aundeley was a verderer of Rockingham Forest at around the same time.

Some forest wardens and justiciars were known for being particularly oppressive. In 1247 the forest of Gillingham came under the control of Robert Passelewe. Under his authority as Chief Justice the courts known as forest eyres were convened more frequently, forest boundaries were scrutinised, and forest officials were fined or removed from their posts. In 1250 he lost the king's favour to be replaced as Chief Justice by Geoffrey of Langley, who is known in Dorset for reafforesting the Forest of Bere. He was particularly resented in the north of England, where he 'miserably impoverished all the forest districts, without reason or pity.'

Alan of Plugenet or Plukenet, a nephew of Robert Walerand, had a family base at Preston (Plucknett) in Somerset. Among his offices was that of constable of Corfe castle, where a tower still bears his shield of arms. He took part in the wars in Wales, Gascony, and Flanders. At his death in 1299 he held substantial lands of his own in Gillingham and Madjeston. During his time of service he received several gifts of timbers from the king and at his decease his son was pardoned debts due to the king 'in consideration of his father's good and praiseworthy service'.

In 1265, and again in the early 1270s, John de la Linde (d.1273) appears as steward or warden of Gillingham Forest. His family seat was at Winterborne Clenston, a village now largely reduced to its church and old manor house. In Dorset his other holdings included the manor of Swyre, where he had been granted a charter to hold an annual fair, and in Somerset he held the manor of Broomfield. He also held manors in Surrey, Sussex, and Lincolnshire. The Inquisition held at his death details that he held not only the bailiwick of the manor of Gillingham, but also that of the forests of Blackmore and Powerstock.[5]

After 1273 the manor and forest of Gillingham were granted to Eleanor, mother of Edward I. In 1299 it was awarded by Edward to the queen, Margaret of Anjou, as a dowry or jointure, and the forest was to be the 'queen's forest' for the following three hundred years. From this time fewer letters and instructions emanated from the royal departments than had been the case previously, and much less is known about the royal officers who had charge of the forest.

Verderers, regarders, and agisters

The Norman kings recognised that the guardianship of their forests needed something more than custody by great men likely to be guided as much by self-interest as by duty to royal rights. This additional oversight was provided through elected, unpaid officers drawn from the knights of the county. Their role was to scrutinise the workings of the forest free of the temptation to benefit from its profits.

One such group was the verderers, initially intended to be twelve individuals 'to keep his [the king's] vert and venison', but in practice of variable number. The office could be held for life provided that the duties were discharged faithfully and well. Verderers supervised and assisted the foresters and keepers in searching out forest offenders and bringing them to the attachment courts. They maintained the rolls of presentments of forest offences for production at the forest eyre, where they could be closely examined regarding the content of their rolls. They also worked alongside the foresters in scrutinising the exercise of customary rights in the forest, such as taking of wood for fuel or the pasturage of animals. The post was exacting, and the verderers could be fined or worse if their evidence was found to be false.[6]

References to Dorset verderers appear in the rolls of the forest eyres for the county, and occasionally in royal instructions. In 1236 Henry III instructed the sheriff of Dorset to direct the verderers and twelve knights of the county to view the condition of the forest, noting how well its vert and venison were cared for. In 1291 the sheriff was directed to find a new verderer for the forest of Gillingham following the decease of one of its number.[7]

A similar group of elected, unpaid knights were the regarders. Their principal purpose was the carrying out of the forest 'regard', conceived as a triennial enquiry into the state of the vert in the forest, and offences against it. Its primary object was to prevent the destruction of the vegetation which afforded food and shelter for the king's deer. By the reign of Henry III regards were held as a preparation for a forthcoming forest eyre.

The subjects of the scrutiny were laid down by law in the 'Articles of the Regard', which stipulated in detail what was to be investigated. The principal items most relevant to Gillingham were assarts, or encroachments on the forest land for enclosure and farming; purprestures, or unauthorised buildings or other features likely to interfere with the movement of the deer; the taking of woods or underwoods for building, fencing, or firewood; and the cutting of woods beyond established rights, known as 'waste'. Other articles of the regarders included enquiry into the unlawful pasturing of animals on the king's grounds 'where no one had common', the condition of the eyries of birds of prey, and the unlawful possession of bows, hounds, or other things likely to damage the deer. In Gillingham, many of these requirements occur as part of the orders for the swanimote courts dating from Elizabethan times (p.52).[8] There are few records of when regarders actually visited Gillingham, but there is one from 1278:

9 The hart (from Cox, The Royal Forests of England).

> The Sheriff of Dorset is to cause a regard to be made of forest before the coming of the justices, so that the regard shall be made before midsummer next.[9]

A third group of unpaid knights were the agisters, who were to supervise the agistment of woods, that is, the driving of swine into the royal forest to feed on the acorns and beech mast. In 1184 Henry II had required the appointment of four knights to oversee this, and to receive and account for any dues received from this 'pannage'. While few mentions of agisters occur in the Gillingham records, references to the value of pannage show that this duty must have been carried out.[10]

All of these unpaid, elected offices would have been unpopular with local knights, since the duties, themselves arduous, involved the enforcement of a system considered hateful and oppressive. Since verderers and regarders were liable for any falseness in their returns, it is not surprising that a knight might try to buy himself out of this duty. By the fourteenth century the system of regards was itself in decline, as was the holding of forest eyres to which they were linked. The duties of regarders came to be more concerned with ensuring that the Crown received its proper dues from the fines paid for trespasses, rather than preventing the trespasses from occurring.

The fee forester and keepers

The forest wardens, while accountable for the forest, often had other Offices and commitments elsewhere, and were not concerned with its

10 The hare (from Cox, The Royal Forests of England).

day-to-day management. This was left to a different official known in Gillingham as the fee forester, or forester of fee. The fee forester, with his deputy keepers or foresters, was the true guardian of the forest, a man well experienced in woodcraft and with detailed knowledge of the beasts of the chase. Unlike the warden, who was appointed only at royal pleasure, the fee forester held his post by the form of tenure known as serjeanty, with a grant of land and a tenement held from the Crown. By the thirteenth century, or possibly much earlier, the post had become hereditary, perhaps an acknowledgment that the specialist knowledge of the forester was best preserved and enhanced by being transmitted from one generation to the next.

The role of the fee forester, which extended to management of the park and forest walks, was the safe keeping of the vert, or green cover of the forest, and its venison. He was expected to work alongside the verderers to arrest and attach offenders for presentment at the next forest eyre. He was to maintain the stocks of deer and the growth of the forest cover in preparation for the royal visits. This included the cutting of timbers in accordance with royal directions, woodland management such as pollarding or coppicing as required, overseeing the permitted killing of deer, and ensuring that deer in the park had the right conditions of shelter and food to breed and thrive. At the occasions of a royal hunt he would help the royal huntsmen to find a suitable quarry for the chase.

The reward for the office, a virgate of land in the Gillingham common fields, together with a dwelling, might seem modest. However, in his position, with only infrequent oversight of his activities by outsiders, it was not difficult for the forester to add various benefits to his office. These could include additional pasture for his animals, unauthorised kills of venison, and timber for the repair of his house and fences. Such acquisitions were regarded by the forest justices and verderers strictly as irregularities, to be enrolled and presented at the courts of forest eyre.

The office of fee forester is first recorded in 1230, but it is likely that the post had its origins in the estate of Edward the huntsman recorded in Domesday Book, where this servant of the king held land amounting to half a virgate (p.7). By the thirteenth century the forester's holding had grown to a virgate, with common pasture for all his animals in the forest and timber for repair of houses or fences. In 1311, at the time of the death of the fee forester, John Joce or Goce, the holding comprised a capital messuage, three virgates of land and four acres of meadow, with rents from four tenants; the 'lop' and 'top' of trees given to others by the king, all windfall wood, unlimited pannage for swine, and pasture for specified numbers of cattle and horses. The forester was to have the left shoulder of every deer taken in the forest and park.[11]

The capital messuage can be readily identified as East Haimes. This was a prominent house which is clearly identified on the 1624 map of the forest. It stood a short distance from the later Windyridge Farm, close to the boundary of the deer park and the Barrow Gate. This was a convenient position which allowed the forester easy access to the park and also allowed him to notice movements of people and livestock around the park's northern edge. A short ride brought the forester to the main gate of the park at King's Court. A house stood at East Haimes until the eighteenth century. While some of the land mentioned may have been close to the house itself, other land was in the common fields of Gillingham manor, where the forester held land on the same terms as other tenants. This included pasturage in the common pastures of the manor for specified numbers of cattle and horses. The 1624 map also marks the position of other lodges or houses of the lesser keepers.

In the thirteenth century the office of fee forester was in the hands of the Joce family. We first hear of this family in 1230, when one John of Winterborne was given temporary custody of the bailiwick while the children of 'Joceus Forester' came of age. The forest eyre of 1246 found that John Joce, the fee forester, had claimed to hold not just his allotted virgate, but also common pasture in the forest for all his animals without limit, as well as husbote and heybote (timbers for repair of houses and fences). The Joce family still held the office and bailiwick in 1299, when John Joce took part in the perambulation of the forest ordered by Edward I.[12]

In 1311 John Joce died, leaving two married daughters who claimed that they should inherit the office with its rights and property. His death produced a disputed situation for a number of years. This arose because the forest was now in the hands of Queen Margaret, who put forward her own nominee to be fee forester, one John Hayward, a steward of the manor. But the court of Chancery found for the Joces, noting that the office had been held by the family for four generations without dispute. Edward II intervened on the side of the Joce family. By this time the elder sister had bought out the younger one and conveyed the office to her husband, William of Bugley. In 1316, two years after his death, it was conveyed to William Haym or Haime.

The Haym family already seem to have been substantial landholders in Gillingham for a number of years. In 1261 a Clement Haym, agister of the forest, died. In 1269 a *Walter in la Hame* appeared in a plea roll of the forest eyre, while in 1274 Adam, Richard, and Walter Haym were recorded as landholders. In 1275 a William Haym was one of a number of people sworn at the 1275 survey or 'extent' of the manor. Around 1300 a William Haym held a virgate of land in the town fields, while Richard (a son), Simon, Michael, and John also held tenements in Gillingham. John, whose tenement was substantial, may also have been a son and later a vicar of the village of Sutton Waldron. The relationships between these various members of the Haym family are not clear, but the frequency of the name at this time indicates a family that would have been well known in and around Gillingham.[13]

The Haym family held the office until 1401, when John Haym was succeeded by members of the Bydyk family, who had become linked to the Hayms by marriage. These were John Bydyk (d.1421) and his father William (d.1430). At his death William Bydyk was lord of the manor of Silton, and also held the 'fee of the bailiwick of the forester of the forest of Gillingham' and various closes and messuages including East Haymes. From the Bydyks the office later passed to the lords of Stourton, by whom it was held in the sixteenth century.[14]

The succession of the office through these families suggests a shift in the role of the fee forester. With the ending of any likelihood of royal hunting in the forest, there was less need for a resident forester with detailed knowledge of game and the pursuit. The routine tasks of finding venison for royal tables could be readily undertaken by the lesser keepers and huntsmen. While the hereditary nature of the office could be maintained, it could now become the property of local gentry families, valued for its forest rights and perquisites. They in turn could appoint lesser men to carry out any actual duties.

The fee forester managed a number of lesser foresters or keepers, one for each of the three walks (Laund or Lawn Walk, Clear Walk, and Woods End Walk), and a keeper of the park. In 1274 John and Richard *Parcere* held parcels in the manor, perhaps indicating the existence of the office of park keeper. Other offices known are those of bowbearer, most likely a deputy keeper, ranger, and wood reeve. Much later records from the sixteenth century show that each of these keepers had specific rights of timber from the forest. The keepers were each entitled annually to take 40 loads of oaken browse and four quarter stubs of trees by virtue of their office.[15] Like the fee forester, the lesser keepers would be men expert in the crafts of the chase of the forest. Besides their woodcraft skills, some keepers would have needed a degree of basic education to read the instructions brought from the sheriff, or to render the accounts of expenses required by the forest wardens.

Forest justice

A function of forest administration was the apprehension of those attempting to deprive the king, or after 1273 the queen, of the exclusive use of the forest. To secure this objective, a system of forest courts existed through which the forest laws could be ruthlessly enforced. The forest courts were an addition to the normal law of the land, intended not to replace the system of manorial courts and assizes, but to supplement it in those areas of the country where forest law was applicable. Inhabitants of such areas might therefore consider themselves to be doubly oppressed, subjected to doing suit of court not only at the courts of the manor, but also at those courts dealing specifically with the trespasses of the forest. The much later writer John Manwood, in his *Treatise of Forest Laws,* makes it clear that

> Offences committed in the Forest are to be tried by the Justices of the Forest, and the Offenders are to be punished by the Forest Laws only; because those laws are appointed for that purpose, and differ from all other laws.[16]

At the local level of the individual forest, justice was normally dispensed through courts known as attachment courts. These dealt with all offences against the vert and venison, and were so called since their principal business was the attachment of offenders to answer for their trespasses in the forest. They could not punish offenders but could enrol them for presentment. This would take place when the attachment court sat as a swanimote, or at the forest eyre of the county. Originally the attachment court was presided over by the warden (or his deputy) and the verderers. This led to extortionate practices and complaints that everywhere the foresters and verderers

> from hatred or otherwise maliciously, that they may extort money from someone, do indict or accuse whom they will, and thereupon do follow grievous Attachments, and the innocent Man is punished, who hath incurred no Fault or Offence at all.

The outcome of this was the Ordinance of the Forest of Edward I (1306), in which he enacted that in future the accusations should be verified by a sworn jury, and that subsequent presentments should be confirmed and sealed by all of the forest officers together.[17]

Another local court of the forest, held less often, was the swanimote. The name swanimote (or swainmote) is derived from an Old English word meaning 'meeting of swineherds', a reference to the concerns of this court with pannage, the grazing of pigs belonging to forest dwellers on the forest acorns and mast. Traditionally the first meeting of the swanimote would be on or around 15 September, when the agisters would witness the entry of pigs into the forest. The second meeting date would be 11 November, when agisters would oversee the driving of pigs from the forest. The third meeting, on or around 10 June, marked the start of the 'fence' month when the deer were expected to be fawning, and the forest must be closed and undisturbed.

In Gillingham, nothing has survived of the forest rolls of the local courts at this time, other than some extracts copied into later documents. It would seem that, as in many smaller forests, the functions of the attachment and swanimote courts were combined into a single court referred to as the woodcourt. It is not known how often this court met, but the traditional requirement was for an attachment court to meet every six weeks and a swanimote court to meet three times a year.

Dating from the end of the sixteenth century, there survives the list of orders of the swanimote court for 'the woods and pastures [of Gillingham Forest] which belong to Queen Elizabeth I.' These orders, which might have been first written out over three hundred years earlier, were intended to regulate how the courts should manage the forest. Before each court met, the foresters and keepers were expected to carry out no fewer than twenty-one different checks on the state of the forest. In content, the orders have their origins in the articles of instruction to regarders, already described above. These included the viewing of recent assarts or purprestures, and of ditches and hedges; the noting of recent evidence of felling, such as cut branches and tree stumps; and viewing the condition of the queen's hedged enclosures. Enquiry was to be made into the holders of bows, greyhounds, and animal traps, and the condition of the eyries or nests of hawks and other hunting birds. A reference to noting the existence of forges and mineral workings might relate to the work of itinerant blacksmiths or charcoal burners, people who would have required quantities of wood for their work.[18] The orders are set out in considerable detail, such as:

> Also the woods of the Queen in the said forest are to be viewed and every stump or root either of oak or beech made since the last court is to be noted and it must be enquired whether the woods are declared of underwood and bushes, and that the decay is to be noted.

Even the humble honey bee did not escape scrutiny:

> Also enquiry must be made of honey if it be in the forest who hath it or of right ought to have it.

A matter of particular concern was the lawing (also hambling or expediation) of dogs, thus

> Also it must be viewed whether dogs be expediated according to the laws, customs, and Assize of the Forest.

This meant cutting off the front claws of the dog's forefeet so that it could not chase the deer. In 1250 it was ordered that the hambling of dogs should be postponed until after Michaelmas, presumably on account of the unnecessary hardship this was causing to the local inhabitants.[19] The foresters themselves were to be scrutinised to see if 'any of them do conceal any trespass of the herbage or of the hunting.'

It is not known for certain where the woodcourts were held. The manorial centre or 'queen's house' was at Thorngrove, but this was on the other side of Gillingham from the forest and may not have been in existence until the fourteenth century. King's Court was on the edge of the forest, but it is unlikely that this royal residence would have been made available for local meetings involving mostly common people. East Haimes, the home of the fee forester, is another possible meeting place. However, the most likely venue for the courts is Motcombe. The name means 'meeting place in a valley', and refers to a point not far from where the ancient Mere to Shaftesbury road crosses a small stream called the Cranborne Lake. This made it an accessible venue for many forest dwellers, and it is likely that the earliest parts of the village grew up around this regular meeting place of the forest courts.

The forest eyre

While minor offences such as the taking of underwood, or the failure to expediate dogs, might be dealt with by fines in the swanimote, more serious offences such as the taking of trees or deer were enrolled for presentment at the next forest eyre. The eyres were the ultimate court in the enforcement of the forest law. Presentment at the forest eyre could result in serious penalties for the trespasser. In Norman times penalties were graduated according to the social position of the offender; a freeman might be fined if he killed a deer, whereas a villein could lose his life for the same offence. The Forest Charter of Henry III in 1217 moderated the harshness of the laws, so that no man could lose his life for taking venison, heavy fines being substituted. Trespass by a forester or verderer could result in removal from office.

In the twelfth century it was required that eyres, together with the regards which preceded them, should be held triennially, but by the middle of the following century the intervals between eyres had been much extended. In Dorset, forest eyres are known for 1224, 1247, 1258, and 1269. In the following century six eyres are known between 1333 and 1371. The Dorset eyres, presided over by the forest justices, were held at Sherborne. In November and December of 1224 forest justices held eyres at Wilton (for Wiltshire), Ilchester (Somerset), and Sherborne (Dorset); the Dorset eyre was on the day following Ilchester, suggesting that the session was expected to take no more than a day.

At the forest eyre of 1269 some 17 individuals were fined for 'offences against the vert', which meant taking timbers from the forest. The fines were of 12d or 2s. All had names suggesting that they lived within the forest or close by. One much higher fine was that of 5s 'for the value of a certain cart and mare.' It was a common practice for a persistent offender to be ordered to surrender his cart and horse; here the offender was being given the option of paying a higher fine to enable him to keep his animal and cart. A further list is of a dozen names of individuals from outside the forest, including several clerics; we read of pleas by the parsons of St. Rowald (Rumbold, Shaftesbury), Donhead, and Gillingham, and by the abbess of Wilton, all for the taking of one or two oaks; and a plea by Robert Wickhampton, Dean of Salisbury Cathedral, 'for two oaks taken in the demesne without warrant'.[20]

By the later thirteenth century the forest eyre was falling out of favour as a means of investigating offences against the vert and venison. Instead, the Crown began to favour direct commissions of enquiry into specific offences, which could be set up and concluded more quickly. Such enquiries could both hear the offence and judge the outcome. In 1307, 1314, and 1316 a number of such commissions were issued to enquire

> touching persons who entered parks of Margaret, queen consort, at …. Gillingham … hunted therein and carried away deer. The like of the same, touching persons who entered the forest of …..Gillingham, hunted therein and carried away deer.[21]

In 1335 Edward III authorised a commission into 'defects in parks and manors' in many places, including Gillingham, no doubt aimed at identifying further trespassers against the royal interest. In 1337 another commission was 'into persons who have enclosed lands and places within the manor of Gillingham, now held by Queen Philippa, whereby she or her tenants are excluded from common of pasture in Stour Provost which she used to have at all times.'[22]

The forest eyre was slow to die and was revived on numerous occasions, usually as a desperate measure to raise revenue from the remaining royal forests. It is unlikely to have troubled too many generations of Gillingham forest dwellers, who by the later medieval centuries had come to regard the taking of the forest timbers as part of their normal way of life.

1 A reconstruction of the royal residence or palace of King's Court, about 1250. The outlines of the boundaries and the entrance are still to be seen. The principal building was the Great Hall. A possible alternative appearance of the hall can be seen bottom right. (Gillingham Museum)

2 The pale or fence of the deer park near Donedge Lodge. Despite the passage of the centuries, some sections of the former park pale are still to be found.

3 Forest boundaries: Motcombe and the forest from Cliffe Wood, on the eastern boundary. The higher slopes of the forest to the east of Motcombe had become well settled by the thirteenth century.

4 Forest boundaries: The ford or bridge at Huntingford or Huntleford gave access to the forest on its western side. The name still recalls the association with hunting of earlier centuries.

5 Forest boundaries: Northwards from Duncliffe towards Motcombe and the forest. Duncliffe Hill was the southernmost boundary point of the Forest of Gillingham.

6 Forest boundaries: Enmore Green and the forest, from Castle Hill in Shaftesbury. The boundary between Shaftesbury and Gillingham forest ran across the foot of the hill. For many centuries Enmore Green was important to Shaftesbury for its springs and wells.

7 Motcombe had become the most populous place in the Forest of Gillingham by the fourteenth century. The base of a medieval stone cross stands outside the church, which was rebuilt in Victorian times

CHAPTER 6

'THE MASTER OF GAME'

Medieval forest law recognised four beasts which constituted the venison of the forest. These were the red deer, the fallow deer, the roe deer, and the boar. The authors of medieval hunting treatises made more subtle distinctions, not all in agreement with each other. The *Book of St. Albans*, written around 1486, distinguishes between the beasts of the venery, identified as the hart (the male red deer), the hare, the boar, and the wolf; and the beasts of the chase, which were the fallow deer (buck and doe), fox, marten, and roe deer. A further group of animals of interest to medieval hunters were the beasts of the warren, which were normally taken to include the coney (rabbit), partridge, pheasant, and fox, and might also include the hare. With the exception of the marten and wolf, all of the above would have been known in the medieval forest of Gillingham.[1]

The red deer and roe deer are native to Britain, and were widespread in medieval times. The red deer, Britain's largest land mammal, was at home in both open grasslands and wooded countryside, and grazed a wide variety of plants from grasses and heathers to trees and shrubs, but today is more confined to the upland parts of Britain such as Exmoor. To medieval huntsmen, the hart or fully grown male deer, was the most challenging and prized of quarries, with the stag, not quite so mature, a close second. The smaller roe deer, 'dainty and spare', was less esteemed, and was considered much easier and less exciting to hunt, and was thought by some to compete with other deer and drive them away. It was, however, valued for the quality of its meat, thought to be 'most wholesome to eat of any other wild beast's flesh.' It was eventually excluded from the protection of the forest law, and this may have contributed to its decline.[2]

The fallow deer is not native to Britain and was introduced by the Normans, becoming a principal species of the deer parks which began to spread across the countryside. With its strong herding instincts, it was well matched to the shorter hunts of the smaller, more confined park spaces, but could also give a good chase in more open country. Although smaller than the hart or hind, the fallow buck and doe produced a good proportion of meat to body weight, and were valuable contributors to royal larders. Because of the different habits of the red and fallow deer, some landowners kept them apart in separate parks. One estimate of stocking ratios in parks gives one fallow deer to five acres, but other estimates suggest much higher densities.[3]

The wild boar was becoming extinct in England by the thirteenth century, but landowners are known to have imported boars for hunting in their parks. There is no record of boars or pigs at Gillingham, but there is no reason to think that they were not hunted there. Boar hunting certainly continued in the area into later centuries. In Cranborne Chase the boar remained protected under the Chase law until the sixteenth century. In 1450 the vicar of Iwerne was prosecuted for killing four swine with his bow at Iwerne Wood. Further north in Wiltshire, boar were hunted in Savernake Forest in 1543, and Charles I endeavoured to re-establish them in the New Forest.[4]

Medieval lords kept many other game animals and birds in their parks. Partridge, swan, pheasant, and peafowl were maintained by nobility from the mid twelfth century, as much to decorate their parks as to provide a source of food. The smaller animals would have been an important nutrition source for the lesser forest dwellers such as the keepers and parkers. At Lodge Farm, Kingston Lacy, the study of a parker's house has revealed bones of peafowl, hare, and rabbit. Rabbits, introduced to Britain soon after the Norman Conquest as a luxury food for the rich, soon escaped their warrens and rapidly proliferated. Rabbiting therefore became thought of as a low activity, not suitable for lords, but was frequently followed by park keepers, and by ladies using ferrets.[5]

Another warren beast, the hare, was considered by some medieval huntsmen to be a beast of some status, largely because of its cleverness at eluding capture. Some hare warrens enjoyed the same level of protection as the beasts of the venison. One such warren was that of the royal manor of Somerton. Here the finding of a dead hare could lead to an attachment of an offender, and appearance at the forest eyre.[6] While not all of these types of game are traceable in Gillingham records, there is no reason to suppose that the forest of

Gillingham and its park were any less diverse in their range of quarries than any other royal forests and parks of the time.

Throughout any forest, the rhythm of the year was dictated by the needs of the game it harboured. A period to be critically observed was the fence month or fawning period for the red and fallow deer, normally taken to be the fortnight on either side of Midsummer day. During this time it was essential that the deer should be as undisturbed as possible, and should have ready access to cover and fodder. To achieve this, activities normally permitted were restricted or prohibited. Other grazing animals such as cattle might be excluded, and rights of way through the forest could be curtailed. Foresters and keepers might spend considerable time in lodges or makeshift huts to make sure that the period was strictly observed. With the end of the fence month up to around September 14 came the time of 'grease' for the harts and bucks, the time when they carried most venison and fat, and when pursuit of them was at its most advantageous. Later in the autumn came the rutting period, another time when access to the forest might be restricted. Hinds and does were often taken over the period between mid-November and the end of February. This might also be a time of hardship for the deer, when a shortage of fodder might result in 'heyning', or the exclusion of other animals, in order to maintain such browse as could be found there exclusively for the deer.[7]

The art of hunting

The knightly and aristocratic classes of medieval England had three consuming passions: warfare, courtly love, and hunting. Hunting with hounds or hawks was considered to be training for war and a rite of passage towards manhood. It was pursued with ardour and ritual, and became an inescapable activity for kings and princes anxious to demonstrate authority and power. The end product of the hunt, a table adorned with venison, proclaimed the high social status of the consumer. Even so, it may have been more important as a special item for feast days than as regular fare.[8]

At the pinnacle of this demonstration of power and status was the pursuit of the hart. Its weight of up to 420lb made it the largest and most mature of the male deers, notably larger than its female, the hind, and esteemed above the less mature stag. The hart was the subject of much religious, artistic, and literary symbolism, and its pursuit was

11 A lymer or bloodhound, with his keeper (from a French hunting treatise)

undertaken in an atmosphere of ceremony and ritual. Of particular mystical significance was the rare white hart. In Celtic times the white hart was considered a harbinger of doom, a living symbol that some moral law had been transgressed. In Arthurian tales its appearance is taken to mean that the knights should embark on a quest. It was later embraced by the Christian church as a symbol of Christ and his presence on earth. In the story of the conversion of St. Eustace, a crucifix is seen between the antlers of a white hart.[9]

In Dorset the Forest of Blackmore has a particular association with the white hart, and indeed became known as the Forest of the White Hart. This followed a story from the time of Henry III in which the king, while out hunting, came across a 'beautiful and goodly white hart', which he spared because of its comeliness. Sometime later the hart was slain by the knight and bailiff of the forest, Thomas de la Lynde. The king was so enraged that the hapless knight and his companions were imprisoned and fined, and he imposed a tax upon the land trodden by the hart. This tax known as the 'white hart silver', was levied in the area for centuries afterwards, although the story itself has often been the subject of doubt.[10]

The Master of Game is a hunting manual of the fifteenth century, but describes practices common in earlier centuries. In it the author describes the precise stages through which the hunting of the hart in the manner known as *par force des chiens* (by force of dogs) is to be achieved. It is not too difficult to see how the practices related in such hunting manuals might have been played out in the Forest of Gillingham at the time of King John or earlier. Once it was known that the king was expected and would wish to hunt, perhaps on the following day, the opening ritual of the sequence would be set in motion. This was the quest, in which the fee forester, or other experienced huntsman, would track down a hart with the aid of a lymer, or scenting hound. The forester would be particularly looking for the largest, strongest beast of the forest. He would bring the information of the hart's location to the *assembly* or gathering of the hunters at King's Court.

At the start of the hunt, perhaps the next morning, the different types of hounds required for each stage of the pursuit would be gathered together. Relays or groups of dogs would then be stationed in positions where the hart might be expected to run.[11] The hunting party, a mix of courtly participants, huntsmen, and dog handlers, some on foot and others mounted, would move out from the encampment around the palace and around the edge of the park into the forest proper. The pursuit would be initiated by running hounds or *chiens de courant*, dogs which could keep to a scent and follow the quarry at speed for considerable distances. Later in the chase, greyhounds were needed, dogs with the power to overhaul the beast and seize it to pull it down. The keeper of a leash of greyhounds was called a fewterer. An *alaunt*, a large hound or mastiff, might be needed to seize the larger or fiercer beasts such as the hart and boar; in illustrations it is often shown muzzled and wearing a spiked collar. These various types of dog were nurtured and trained by a team of keepers or berners led by the king's master huntsman (Figure 12).[12]

12 *Berners with their hounds (from Cox, The Royal Forests of England).*

The *Master of Game* relates that, once the hart was found and the hounds were in position, the 'unharbouring' and chase would begin. Throughout the hunt, horns would be used to signal the progress of the chase and the hart's location. During the day the entire populations of Gillingham, Motcombe, and Shaftesbury would know from the notes of the horns how the chase was progressing. In view of the stamina of the hart and its ability to run long distances before exhaustion, the pursuit might take the party out of the forest of Gillingham and across Duncliffe into the forest of Blackmore around Marnhull and beyond.

Eventually the hart would be trapped by the pack of hounds and would turn 'at bay' before being ritually slain by the sword of the king or his chief huntsman. This was followed by the 'unmaking', the cutting up of the hart at the site of the kill to a recognised, carefully ordered pattern. At the unmaking the fee forester would claim his right shoulder of the beast, and other foreparts of the carcass might be set aside for the other forest keepers involved. Then came the *curee*, the rewarding of the hounds with those parts of the carcass specifically set aside for each type of hound. The journey back was equally ceremonial, with horns blowing at the gate of King's Court announcing the royal arrival. The trophies of the hunt, which might also include some of the lesser kills of the day, could be set out in the great hall for all to see.[13]

Bow and stable

Pursuits of smaller deer, such as stags, hinds and fallow deer followed a similar pattern but were more perfunctory and concluded with less flourish. For the fallow deer an altogether less energetic, more relaxed type of hunting was followed, known as bow and stable hunting. In this method, teams of hunters and their dogs would drive groups of deer towards pre-prepared 'stations' or 'stables' where bowmen would be waiting, ready to despatch the quarries. Wounded deer could then be pursued and finished off by the hounds. This was the most common type of hunting found in Britain before the coming of the Normans, and would have been the method employed by Alfred in his Wessex forests.[14] After the Conquest it became the most usual mode of hunting in enclosed areas and deer parks, where the woodland could be managed to provide glades and spaces along which the pursued deer could be driven.

Bow and stable hunts could be of varying scale. At its most basic, deer could be taken by just two or three mounted hunters, the same number of bowmen, and a single dog. The group would make clever use of the cover and the wind direction to startle the deer into moving into the range of the stationary archers. The dog would be used to pursue any wounded beasts. Other bow and stable hunts might be large affairs with many huntsmen, beaters, and dogs. In this version of events the riders and beaters would endeavour to trap a group of deer using the terrain of the forest to drive the quarries down wooded valleys or gullies from which there was no escape; at the end would be the waiting bowmen.[15]

In this more 'gentle' mode of hunting, regarded as pastime rather than rigorous exercise, women might feel encouraged to participate, using the smaller crossbow specially designed for ladies:

> The kingdom of England is adorned with fine hunting, for there is such abundance of parks, full of venison ... and when the ladies resort there for entertainment they take singular joy in shooting with the bow and killing these beasts.[16]

As on the ground, so in the air. The courtly classes of society were expected to be as familiar with the art of hunting with hawks as they were with hounds. This was a less formal activity than deer hunting and was not regulated by the forest laws. Its pursuit was not confined to forests, and might take place whenever groups of enthusiastic falcon owners found themselves outdoors together. Falconry was normally a more leisurely alternative to deer hunting, but when undertaken by the royal retinue might become a more organised affair. A large party of nobles and royalty, each with a hawk on his or her wrist and accompanied by retriever dogs, might spend a day taking it in turns to down partridge, pheasant, lark, duck, heron, or crane, to the acclaim of the other members of the party. At the end of a good day they would return to the castle or hunting lodge with bags overflowing with game.

Unlike deer hunting, falconry was not limited to the princely or knightly classes and could embrace participants from further down the social scale, such as squires, clerics, and burgesses. An essential requirement was willingness to acquire a sparrowhawk or other bird of prey, and dedicate several weeks of care and discipline to its training. The hawking party was a convivial affair, all about good company and an enjoyable time. One writer on the subject has written:

> One needs a stout, steady-going horse, another in reserve, and four spaniels to quest and retrieve ... the other requirements are good country and good company; lone hawking is poor sport. The company should be blithe and beautiful: knights and squires ... ladies and maidens ... each bearing a

sparrowhawk ... in a friendly and flirtatious competitive spirit, until the game bags are filled.[17]

One can only suppose that the forests and countryside around medieval Gillingham and Shaftesbury would be as appealing as anywhere to the gentlefolk of the royal retinues and of the neighbouring towns and manors. For any ladies of the royal party not wishing to expose their falconry skills to the comments of the menfolk, the seclusion of the deer park offered an alternative venue for practising the art away from prying eyes and unwanted advances.

13 A gentleman out hawking (from the Luttrell Psalter).

Deer for the royal larders

In July of 1313 John Lovel, a king's huntsman, was despatched to a number of forests in the south of England, along with two berners (dog handlers), four fewterers (greyhound handlers), 24 running dogs, a bercelet (lymer or scenting hound), and 16 greyhounds. They were to take and deliver for the king 96 harts and bucks. At the same time, a further two huntsmen, together with their hunting packs, were sent to forests in the midlands to bring back 108 harts, bucks, and hinds. These were much travelled men who had already been on hunting expeditions earlier in the year, and would make more hunting trips before the year was out. In December Lovel was in Somerset and Dorset with his party, but it is not specified which forests he visited. The sheriff of the two counties was instructed to pay his wages and costs at a rate of 12d a day for Lovel himself, 2d daily for his fewterers, and 1½d for his berners. Each of the dogs was to have daily costs of ½d. These expenses would have been deducted from the

'The Master of Game'

farm of the county, the amount of revenue from the county which was submitted by the sheriff to the king.[18]

These trips were not about the pursuit of the deer in the ritualised, *par force* manner related above, as a royal sport, but were to do with the much more workaday task of bringing back meat for the royal larders. Out in the forests the deer would be tracked down methodically and finished off using bow and stable tactics, the aim being to collect as many carcasses as quickly as possible from any expedition. Local foresters would be recruited to help with the drive, and the sheriff would be able to provide any additional bowmen needed for the kill. Once the carcasses were secured, they were handed over to a lardener, whose job it was to salt and keep secure the slaughtered deer. The sheriff or forest warden was expected to provide the salt and barrels needed for transit, and arrange for delivery to wherever required.[19]

While Gillingham does not specifically feature on Lovel's 1313 itineraries, there are several references from the previous century to the king's huntsmen visiting the Gillingham forest. In July 1241 the forest warden, Robert of Aundeley, received a note that he should expect the king's huntsman, William Lovel, to visit the forest to take 20 bucks. The warden was to find the salt needed and keep the carcasses until further notice. In November of 1244 Robert des Mares and James Hoese, king's yeomen, visited Gillingham to take 30 does and five hinds, to be salted and carried to Westminster for the king's larder. In July of 1246 Mares and another king's man appeared in Gillingham to take 10 deer. Ten does were required in December of 1252, no doubt a contribution to a royal Christmas celebration. In November of 1256 the king sent his huntsman Robert of Admundestrop to take 24 deer 'as provision for the approaching feast of Christmas.' In August of 1258 Henry of Candeur and John Lovel took 24 deer from the forest 'for the king's service.' This John Lovel was clearly not the one mentioned already in 1313, but the persistence of the Lovel name in these references might suggest that the huntsman appointments ran in families or were hereditary.[20]

The evidence of royal huntsmen, travelling regularly on commissions to return with quantities of game for the royal larder, suggests a changing pattern of deer pursuit in the royal forests. At the time of John and earlier, the huntsman stayed close to the royal entourage, since the demand for meat was where the court was travelling, and much of the need would be satisfied by regular forays into the forests along the journey. As the court became less itinerant, the royal larder became more Westminster-based, and the venison requirement could only be filled from the sorts of hunting trips described above. It has been calculated that in the mid-thirteenth century the king was taking around 600 deer a year from his forests and parks. At Gillingham the deer contributing to the larder could be most readily taken from within the park, the rest of the forest acting as a reservoir from which the park stocks could be replenished.

Deer as gifts

The use of the royal forests as a larder was supplemented by their function as a source of gifts by which kings could reward officials and followers for their services and loyalty. Dispensation of such largesse was an essential part of the demonstration of royal power and authority. It was expected that favours and service showing allegiance to the king would be rewarded with gifts such as deer, hunting birds, hunting rights in the royal forests, or timber for building works; or in the case of long service and dedication in battle, with grants of property or manors. The king might himself offer such endowments as a way of seeking loyalty and trust. In the case of gifts of deer, the bounty might take the form of carcasses, or live deer for slaughtering later. With the growth in popularity of baronial deer parks, numbers of live fallow deer might be donated as a way of restocking parks which had become depleted of game.[21]

Examples of all these types of gift can be found in the royal rolls relating to Gillingham at the time of Henry III and Edward I. Several gifts of deer were made to members of the king's family. Stephen Longespee (d.1260) received five bucks in 1258; he was the son of William Longespee, John's half-brother, and had served Henry as seneschal or steward of Gascony (at that time an English possession) and Justiciar of Ireland. Stephen's mother, Ela, countess of Salisbury, (d.1261) was given three carcasses of venison in 1258; she had been the founder of Lacock Abbey, to which she retired after she had been widowed. Another Longespee, Nicholas, Bishop of Salisbury, had six bucks in 1292.[22] A half-brother of Henry, Aymer de Valence (c.1222-1260) had 10 bucks from the forest in 1251, and in 1253 had a further five bucks and 15 does for his deer park at East Knoyle. He was Bishop of Winchester at this time and was very unpopular, becoming a focus of the discontent with Henry's foreign favourites. Another brother of the king, Richard earl of Cornwall, in 1242 had 10 bucks and 20 does from the forest for his deer park at Corsham; he was a powerful noble who later built the castle at Mere and laid out a further deer park adjoining the forest of Gillingham.

A second group of recipients for the king's deer were officers who had served the king well at war and in administration. Luke of Drumare was a household knight of John in 1215-16 and was granted the manor of Basingstoke in 1217. He received pairs of deer in each of 1231, 1239, and 1240. Nicholas of Molis or Moels held several offices from Henry III, being at times seneschal of Gascony, governor of the

14 The end of the hunt: the unmaking of the hart (from Cox, The Royal Forests of England).

Channel Isles, and constable of castles including Sherborne, Corfe, and Windsor. He had three carcasses in 1257. Eustace of Hache or Hatch in Wiltshire is thought to have risen from being a menial servant in the household of Edward I to become a knight distinguished in the wars in Scotland; he had six bucks as a gift in 1292, and later became a baron and member for parliament. On the more civil side, William of Ralegh (d.1250) was a distinguished judge, administrator, and king's adviser, becoming bishop of Norwich and then of Winchester; he had two carcasses of venison in 1231.

Elias de Rabayn (d.1285) was a Poitevin favourite of Henry III, holding many offices and gaining many rewards from the king. He had 30 bucks and does from the forest of Gillingham in 1251, a further four in 1252, and three carcasses of venison in 1257. A particularly interesting example of royal service is that of Iterius of Engolisama (or Angouleme, in Gascony). He was a cleric who went abroad frequently in the royal service. In 1282 he was sent by Edward I to collect £1000 from the constable of Bordeaux. He was then joined by two other royal envoys, and together they travelled to Castile and Aragon to buy horses for the king. He is recorded as having 10 bucks from Gillingham in 1292. He was archdeacon of Bath in 1295-6, and canon of Dublin cathedral in 1297. For some of the recipients of the royal bounty, the gifts from Gillingham were only a fraction of what they received from all the king's parks and forests.

Some deer were destined for clerical destinations. In 1258 Agnes Ferrers, abbess of Shaftesbury, was given five carcasses for a feast day. In the same year the abbess of Elstow in Bedfordshire, Aubree de Fecamp, was given three carcasses. Elstow was well known, and by many people frowned upon, for its secular attitudes and behaviour, the nuns often having their own households within the abbey and enjoying a social life little different from that outside the walls.

The gifts of deer from Gillingham seem to have been largely targeted at people within or close to the royal circle of courtiers, officials, and distinguished soldiers. Few local figures seem to have been honoured in this way. One was Alan de Plugenet of Somerset, who at one time was warden of Gillingham forest. In 1271 he was given two carcasses of venison, and his wife received six bucks in 1292. Another, rewarded in a different way, was Robert de Mucegros of (Charlton) Musgrove. He had no gifts of deer, but in 1243 was allowed free warren over the forest of Gillingham, with the right to take fox and hare. It seems that the association of deer with power, authority, and status was something that was not easily relaxed.

CHAPTER 7

THE GREEN COVER

At Kingsettle, on the eastern edge of the parish of Motcombe, half a mile or so north of Shaftesbury, a small piece of woodland clings to the edge of the Greensand ridge which overlooks the Vale of Blackmore. This wood, now of around 20 hectares, is prominently shown on the 1624 map of the forest, and appears to be a vestige of the Gillingham forest of medieval times. The wood is now designated as Ancient Woodland, and consists of oak, ash, sycamore, and beech, with more recent plantings of coniferous trees. Although all the plantings seen today are of later centuries, a walk through the wood to look across towards Motcombe and Gillingham can give a feel of what much of the forest of Gillingham looked like in earlier times.

Woodland of some sort may have covered Kingsettle continuously since the end of the Ice Age some ten thousand years ago. At that time the great primeval forest or 'wildwood' covered three-quarters of Britain, but by the time of Domesday Book this had shrunk to around fifteen per cent. In the south-west the remains of the wildwood were represented by great forests such as Selwood, which stretched across Somerset and into Dorset and Wiltshire. The forest of Gillingham, protected by the royal forest laws, was one of the longer surviving parts of the old forest of Selwood. A local story is that at one time the woodland was so dense that a red squirrel could move from Gillingham to Shaftesbury and on to Mere without having to descend to the ground.

Iron Age and Roman farmers in Wiltshire and Dorset had concentrated their efforts on the lighter soils derived from chalk and limestones. Around their villages and settlements they set out small fields, and also grazed sheep and cattle which steadily turned the vegetation into pasture and prevented its reversion to forest.[1] By contrast, the heavier clay soils of the Vale of Blackmore around Gillingham attracted settlement more slowly, allowing the natural woodland to flourish longer. Here the woodland cover was dominated by the broad-leaved oak forest, and the elm. Without human intervention, individual oaks could grow to well over 30 metres and might flourish for hundreds of years. Where a tree died or was felled, the dark forest cover would be opened up to the sunlight, giving rise to a rich ground flora of flowering plants and shrubs. The ancient or dead wood in turn provided the material for a rich growth of lichens and fungi.

The green cover of the medieval forest was known as the *vert*, and enjoyed the same level of protection under the forest law as the venison. It was important to maintain a supply of food for the deer, which preferred to browse on the trees rather than eat the underlying grass. A mature red deer might be expected to eat the equivalent of its own body weight in fresh forage every 10 to14 days. It was a task of the foresters and keepers to ensure that the deer had a continuous supply of browsewood throughout the year. This could be a problem in winter, when extra feed might have to be provided from hay, from hollins (browsewood cut from holly bushes), or from hags (enclosure set aside to supply browsewood).[2] Other demands on the forest cover came from the need for mature oak timbers for building works not just in the forest but in royal properties elsewhere; and to maintain a supply of smaller timbers for fences, gates, hurdles, wattle, and thatching. While oak is the timber most frequently mentioned in official records for major building works, other species such as elm and ash may have found a use for lesser purposes such as flooring and smaller buildings.

Coppicing and pollarding were the most common methods of woodland management. Coppicing involved creating compartments in the forest protected from deer grazing by banks, ditches, or fences. Once the trees had been cut down to a low level the coppice shoots would provide a regular supply of posts, staves, and spars. Pollarded oak trees had their branches severed to a few feet above ground level. This might take place every eight to fifteen years. The resultant new growth of long, thin branches would produce thick poles and timbers which could be sawn into planks for building work.

Some parts of the forest would have been subjected to more woodland management than others, with many areas being left entirely in their natural state. The foresters would have been most active in areas around the park and King's Court. Here they would have created secluded glades where the deer could find shade in the summer, protection during fawning time in the 'fence month', and shelter in

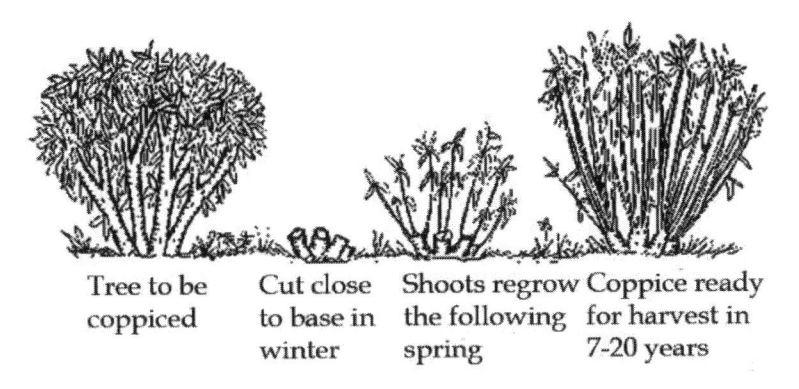

| Tree to be coppiced | Cut close to base in winter | Shoots regrow the following spring | Coppice ready for harvest in 7-20 years |

15 Coppicing in the forest.

the winter. Elsewhere grassy clearances or 'lawns' would be provided, where deer could be readily found and easily taken when venison was required.

Fallen trees and dead wood, known as *rothers*, were of considerable value and were reserved for the king's use. From time to time an enquiry might be made of the value of fallen wood. In December of 1222 the king directed the warden, John of Monmouth, together with the verderers and foresters, to see that all fallen wood was viewed and valued, and that none should be removed until this was done. In the standing orders of the swanimote courts, 'every stump or root ... made since the last court' was to be noted in order to identify possible thefts, along with places where the wood had decayed or deteriorated into underwood.[3]

Commons and pasturage

The livelihoods of the inhabitants of Gillingham manor were closely bound up with the forest. All the tenants of the manor had common rights within the forest, particularly the right to take wood for buildings and fences, and the right to graze cattle. Sheep were generally excluded under forest law because it was believed that their grazing habits competed with those of the deer. For the majority of tenants, the grazing rights appear to have been unstinted, which means that they were not regulated by the number of animals they kept in their own fields and enclosures. This was particularly important to poorer tenants and cottagers, who came to depend heavily on the common pasturage for the few animals they could afford to keep. Similar rights were enjoyed by several smaller places within the manor, and some others outside it. Along its northern edge the forest ground rose gently towards the adjoining manor of Mere, whose inhabitants also claimed common rights within the forest. The demands on forest rights, particularly pasturage, therefore came to be considerable, giving rise to regular movements of livestock across the forest boundaries.

Over time the continuous grazing would steadily transform the character of the vert, preventing renewal of the oaks and creating woodland glades and clearances with a profuse growth of grasses and herbs. It was in the interests of the keepers and foresters to limit or control grazing since it gradually reduced the amount of cover for the deer. From time to time enquiries or inquisitions were held to confirm the rights of the tenants, sometimes following attempts to restrict them by the forest officials. In September 1278 such an enquiry was held at Shaftesbury before Walter Scammel (Dean of Salisbury Cathedral), William de Wyntershulle, and fourteen sworn knights and others. This followed complaints by the tenants against Guy of Taunton, the queen mother's bailiff, that he was denying them their rights of common. The enquiry found that the men of Gillingham 'were accustomed to have common within the covert of the forest of Gillingham with all their animals, except, pigs, sheep, and goats, everywhere except in the region of Marleghe and the park, but by what warrant or from what time the jurors know not.' An exception was in the fence month, a time set aside when the deer could fawn without disturbance.[4]

The 1278 enquiry found that these pasturage rights were shared by the inhabitants of Milton, Pierston, Wyndlam (near Silton), Madjeston, and Wyke, by the tenants of the prior of Montacute at Ham, and those of Walter of Winterborne. The right also extended to the men of Mere 'on the north side of the park'. Mere was also a manor of the Crown as part of the Duchy of Cornwall, and the arrangement perhaps reflects a once poorly defined boundary between the commons of Mere and those of Gillingham.

Crown and tenants could find themselves in dispute over grazing animals within the park. Despite the existence of the park fence, tenants somehow still managed to find ways into the park for their animals. In 1228 the grazing rights of the Gillingham men within the forest were confirmed, provided that they stayed outside the deer leap and did not enter the park. The problem clearly persisted, for the following year any livestock found in the park were ordered to be removed and used for fattening for the royal larder.[5] Another problem was the grazing of animals in the forest by manors other than Gillingham. In 1237 it was ordered that swine found in the forest which belonged to the Bishop of Winchester should be returned to him; the most likely source of the intrusion was from the Bishop's nearby manor of Knoyle.[6]

The forest was also a source of pasturage for the king's or queen's own cattle. With the decline of King's Court, the centre of manorial activity shifted to a domainal centre at Thorngrove, outside the forest on the western side of Gillingham. Here a farm was maintained, known as the Queen's House. From here cattle were driven into the forest, where they were given some priority over the cattle of tenants. In 1251 it was ordered that, because of the damage done to the herbage by the king's oxen, other beasts were to be pastured outside the bounds of the forest, in and around the town instead, a requirement which no doubt caused the tenants some hardship.[7] Some tracts of forest were set aside specifically for the royal oxen, one of which is so described:

> towards St Edwards [ie Shaftesbury] through Fernbroc as far as the park ditch runs to the Leveresmore brook, and by that brook as far as Blindelack, and thence as far as the headland above Alemore, and from Alemore across Geldenemore and thence to Falemore through the upper bend of Falemore as far as Farnbroc through the Portway; that woodland to be held for the aforesaid stock according to bounds indicated.[8]

The references to Fernbrook and Blindlake would suggest a tract of country lying to the south of the park, bordered by these streams, and extending up on to Duncliffe Hill. Reference on the 1624 map to a 'Queen's green' on the slope below Duncliffe might appear to confirm this. The name 'Portway' is a reference to the Shaftesbury to Sherborne Road, or Sherborne Causeway. We might envisage the queen's cattle being driven from the Gillingham demesne farm at Thorngrove across the town, and along the Shaftesbury road and Lox Lane to pasture grounds on the Duncliffe slopes. Some confirmation of this appears to come from the rent roll of 1300, where John of

Wyke is allowed 'grazing on the *Donne*, for his own cattle in common with the Queen's cattle'.

In 1276 the rights of the church in the forest were brought to the royal notice. Nicholas of Cranford, rector of the church at Gillingham, claimed that the tithes of produce from places within the forest should be awarded to the church 'according to the form of the apostolic instruction', since the forest was within the parish of Gillingham. He also reminded the king that he had petitioned him on this matter when he had been at Gillingham for Christmas. As an alternative he requested compensation for the tithes, perhaps in land. Cranford also raised the issue of the lawn or pasture named Merleye (Marleghe or Marley), which had belonged to the church and provided tithes of corn or herbage until it was included in the forest at the time of John Biset (Chief Justice of forests 1238-41). He asked that he might have his rights of grazing in Marley and other pastures restored to him.[9]

A particularly important right was that of pannage, the right to browse pigs on the acorns of the forest oaks. Like the grazing of cattle, this right was open to all inhabitants of the manor. It was highly valued by the cottagers and poorer people, who would look forward to the start of the pannage period with anticipation. The Customary of the manor, a compilation of manor regulations and customs, prescribed that between Michaelmas (September 29) and St. Martin's day (November 11), the time of greatest mast or acorn fall, all tenants were allowed free pannage for one pig aged one year or more. Other pigs so grazed had to be paid for at the rate of 2d per yearling pig and 1d for other pigs. Pigs found in the forest at any other times for more than a day and a night were to be forfeited to the crown.[10] The oversight of the pannage was normally carried out by the group of knights known as agisters, whose task was to supervise the driving of swine into the royal forest to feed on the acorn mast, and to account for any dues received.

Timber for royal building

Between 1234 and 1297 the Close and Liberate Rolls record the taking of over 900 oak trees from the forest. These were the timbers for which the felling was officially sanctioned, but there is evidence of many other trees felled. At the Dorset forest eyre of 1269 some 17 people from in or around the forest were fined for offences against the vert, or the taking of timbers, since the last eyre (1257). A further dozen people from outside the forest had between them taken 21 oaks. Since the last eyre, 407 oaks from the forest had been given away for which the warrants could be produced. 238 oaks had been used in the repair of the royal houses at Gillingham, and a further 12 for the works at the castle at Sherborne. A further item was the sale of 200 oaks on the king's behalf by one Adam de Grenville. 131 horse loads of dead trees had been delivered to the abbess of Shaftesbury.[11]

The taking of timbers at these rates, whether authorised or otherwise, undoubtedly accelerated the decline of the forest as a woodland. Orders for the taking of timber sometimes specify that the oaks should be taken from the park, or from outside the park in the main forest. It is likely to have been the park which could supply the tallest, largest oaks suitable for major building works, such as crucks and trusses for roofs. Here, regular management and clearances of smaller trees, undertaken to create glades and spaces to enhance the hunting, would have allowed oak trees to mature with diminished competition from other species. The less accessible parts of the forest, where woodland was less managed, are more likely to have been associated with the shorter growth more suitable for smaller lengths such as posts and poles.

A principal use of Gillingham timbers was for the various royal building works undertaken in and around Dorset. At Gillingham a major rebuilding of King's Court took place between 1250 and 1252, and has been fully described in an earlier chapter. The orders for 270 oak trees from the forest required for the work are traceable in five authorisations in the Close Rolls, one of which reads:

> Nov. 6 1251 John of Venois [Vernun], seneschal of the forest, is directed to cause those in charge of the king's works at Gillingham to be supplied with 80 logs, viz. 40 from the park at Gillingham and 40 from the woods outside to enable the said works to be proceeded with.

The number of oaks taken would have been recorded on a tally stick, one half of which was retained by the local official and the other half sent to the Exchequer as proof of the number of trees felled. There is no record in the Close Rolls of the 238 oaks recorded in the eyre roll of 1269, suggesting that many more trees were used for repairs and building works than appear in the royal rolls.[12]

This was a time when John and Henry III had embarked on ambitious, expensive projects to rebuild and refurbish their many royal castles, residences and lodges, and Gillingham timbers were used to supply other royal manors and buildings, some nearby, others more distant. The references in the Close and Liberate Rolls to the use of Gillingham oaks in these buildings are as follows:

> *Camel*, king's houses, six curved timbers (1229)
> *Corfe castle*, joists for keep (1234), 4 oaks for repairs (1238), 74 oaks viz. 24 for beams for king's chamber, 50 for upper rooms at gate (1244); 60 oaks (1248); 40 oaks for bridge (1248); 20 oak trunks for repair of bridge (1256)
> *Clarendon*, 10 logs for posts and great timbers and 40,000 shingles for roof (1244); 10,000 shingles (1246)
> *Salisbury castle* (Old Sarum), 6 oaks for shingles for roof of outbuildings (1246)
> *Sherborne castle*, 20 oaks for shingles for king's lodgings (1253); 10 oaks for repair of domestic quarters (1256); timber for works (1258); 6 oaks (1268)
> *Dorchester castle*, 6 oaks for chapel (1252, with Corfe).[13]

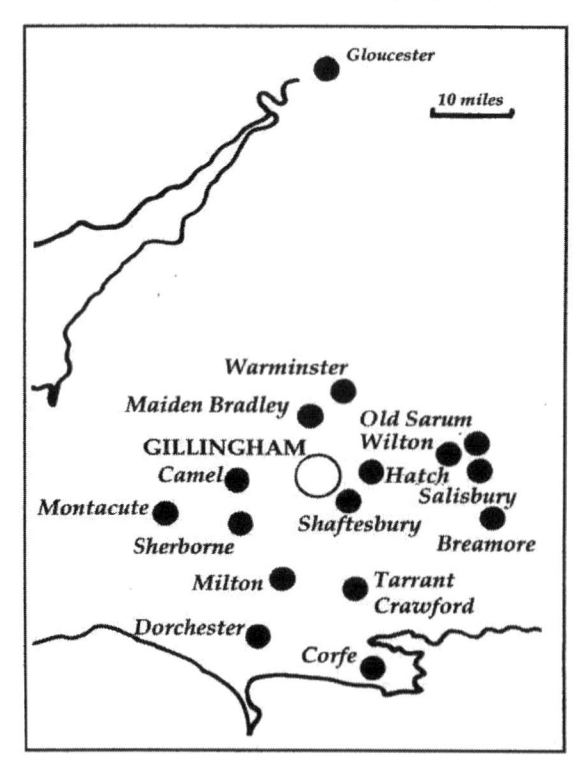

16 Places where oaks from Gillingham are known to have been used.

Sherborne castle, 12 miles from Gillingham, had been first built by Roger de Caen, Bishop of Salisbury, and already extensively altered and repaired by John and Henry III. The castle at Salisbury which had shingles for its outbuildings in 1245 was at Old Sarum, two miles

The Green Cover

north of the city. It is not known which of its several buildings received the shingles, but it is known that the hall was repaired in 1247. East of Salisbury was the hunting lodge at Clarendon, transformed into a palace by Henry II and Henry III, but now long disappeared. It was the residence from which Henry II issued his Assizes and Constitutions of Clarendon, a major codification of English medieval law in the 1160s. Gillingham's contribution to this structure, almost a century later in 1244, was to supply many shingles for the roofing work. Gillingham oaks found their way to the castle at Dorchester, later demolished to become the site of the Dorchester prison. The chapel is likely to have been one of the royal rooms within the castle.

Much less is known about the destination of 'six curved timbers for the works on the king's houses at Camel' of 1229. The manor of Camel in Somerset had been acquired by the king a year earlier from Hubert de Burgh, a major figure at the court of Henry III and his father John. The requirement of curved timbers might suggest their use for the roof of a great hall, but the location of the building concerned cannot be clearly identified.

Corfe castle, 29 miles from Gillingham, was the largest single user of oaks outside Gillingham. The earliest part had been built by William the Conqueror, but was rebuilt and much extended by John. The wood from Gillingham appears to have been used in several parts of the castle, notably the floor or joists of the keep, drawbridge, king's

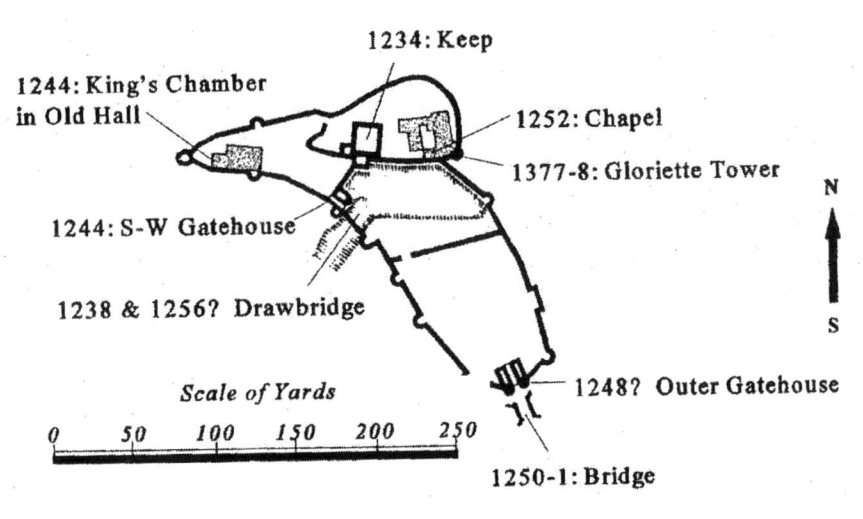

17 Parts of Corfe Castle where oaks from Gillingham are thought to have been used.

chamber, gatehouse, and chapel. In 1377 more Gillingham timbers were used in the construction of the Gloriette tower.

The steady requirements for oak timbers must have meant that regular movements of felled trees were taking place in and around the forest. The heavy haulage of large orders, each comprising several entire trunks, required considerable resources and organisation. By the thirteenth century, long horse-drawn carts were beginning to replace oxen-pulled wains, since they were quicker and could work for longer. Oxen might still be required on steep hills. The routes out of the forest would have taken the timber hauliers on to the higher ground of the chalk downs, where they could follow the ancient trackways and ridgeways for many miles across the open downlands. The haulage teams could expect to make no more than twelve miles a day, perhaps much less in wet, muddy conditions.

The transportation of timbers from the forest to distant places under such conditions was not without its problems. In August of 1250, 40 oaks were ordered for the repair of the bridge at Corfe. In July of the following year, the timber was found to be lying on the hillside at Melbury outside Shaftesbury, and it was instructed that it should be moved without delay. We can envisage that the wood had been loaded on to long ox carts which had failed on the steep gradient at Melbury Abbas, perhaps because of bad weather or the muddy trackways. The load had been spilled, and the task abandoned. In 1256 more oaks were ordered to repair the bridge, perhaps suggesting that the dumped load may never have reached its destination.

Timbers as spiritual gifts

Further numbers of Gillingham trees were distributed as gifts. Gifts of venison (Chapter 5) had largely been to royal officers and favourites, but the main recipients of timbers were some of the many religious houses of Wessex. Gifts to cathedrals, abbeys, and priories were an important instrument of policy for medieval kings. Besides demonstrating spiritual commitment and helping to secure salvation for the royal family, gifts could be used to ensure support amongst the powerful and wealthy ecclesiastics, many of whom had a political and military importance to rival that of the temporal nobility.

Not surprisingly, the nearby and important Abbess of Shaftesbury was favoured with gifts of timber for the nunnery, making the house the fourth largest single user of Gillingham timbers. John gave the

Abbess Marie, who was his great aunt, the right to collect two loads of brushwood daily from the forest. However, Henry III had little regard for one of her successors, Agnes de Ferrers, and she had to fight for this right against the obstructive wardens of the forest. In July 1250 the warden John of Vernun

> is directed to permit the Abbess and convent of St. Edward of Shaftesbury to have two sumpter [ie pack] horses in the forest for the collection of dead wood for a period of 15 days until next Michaelmas, as they have been accustomed to have.

The warden may not have taken this instruction very seriously, for on October 28 the order was repeated, this time specifying the quantity of 200 loads of firewood. The order had to be repeated again in January of 1251, and yet again in April, when the allocation was described as four loads per day. In August of 1251 five timber trees were given by the king, and 20 oaks were requested in 1253. In 1272 15 oak trees were needed for the repair of the nuns' dormitory, and a further 20 in January of 1276. The last-named order may have originated as the result of a visit by Edward I, when he spent the pre-Christmas period at Shaftesbury before spending Christmas Day at King's Court.[14]

Other gifts of timber were to religious houses at some distance from Gillingham. Those recorded in the Close Rolls can be summarised:

> *Salisbury*, house of Friars Minor, shingles for roofing (1233)
> *Salisbury Cathedral*, 50 tree trunks (1234)
> *Maiden Bradley* priory, 4 posts for church tower (1245)
> *Tarrant Crawford* nunnery, 3 oaks for choir stalls (1247) and 10 oaks (1252)
> *Wilton*, Dominican friary, 15 oaks and loppings for new house (1255), 10 oaks (1258), 6 oaks for cloister (1271)
> *Gloucester*, Dominican friary, 5 oaks with loppings (1256), 10 oaks (1257)
> *Breamore* priory, 7 oaks (1268)
> *Milton* abbey, 10 oaks (1271)
> *Montacute* priory, 12 oaks (1297).

Royal generosity was often extensive towards the new monastic orders of the thirteenth century. The Salisbury house of the Friars Minor (Franciscan or Greyfriars) was founded soon after the cathedral and town were built around 1229. It received substantial gifts of royal timbers, mainly from the Wiltshire forests, the Gillingham contribution being towards its roofing. The house stood on a site east of the cathedral close to the later Exeter street.[15] In Wilton a

Dominican friary was established in 1245 in what is now West Street. Despite grants of timbers from neighbouring royal forests, it failed to take permanent root and eventually moved to Fisherton, outside Salisbury, leaving the Wilton house as a cell.[16] Maiden Bradley priory had been founded in 1164 as a leper hospital for women, but by John's reign had become a priory with Austin canons; the site of the house survives at Priory Farm north of the village.[17] The nunnery at Tarrant Crawford had the patronage of Eleanor of Provence, wife to Henry III, and a substantial new church was built in the 1240s. It was the choir stalls in this church for which the Gillingham oaks were needed; the 10 oaks of 1252 may have been used in the construction of the nunnery living quarters.

Milton abbey had been founded in Saxon times around 935, but was extensively rebuilt in the twelfth century. This had been completed by the time of the Gillingham oaks in 1271, so the purpose of this gift is not known. The Cluniac priory of Montacute in Somerset had already been linked with Gillingham long before the time of the gift of 12 oaks through the grant of 'the land of Gillingham which is called Ham' to the monks in the time of Henry I. Once again, the use of the Gillingham oaks at Montacute is not specified. The priory at Breamore in Hampshire had been founded by Austin canons during the reign of Henry I; at the time of the gift from Gillingham the roof was in need of repairs, and some years later in 1278 the priory was still petitioning the king for the timbers needed.[18]

There are only two instances of where Gillingham timbers may possibly still be found within existing structures. In 1234 50 tree trunks were ordered for the construction of Salisbury Cathedral, most likely for the roof. Parts of the original roof structure such as the north-east transept are still to be found, but it is not known if these included Gillingham timbers, or if the Gillingham oaks were used in the roof at all. The other, and more promising candidate for the survival of Gillingham timbers, is the most distant from Gillingham. The building of the friary of the Dominicans or Black Friars at Gloucester, nearly sixty miles from the forest, began in 1239. This is one of only three Dominican friaries to survive in the whole country. The roof timbers of the south range have been dated by dendrochronology to a felling period between 1230 and 1269. This is close to the 1256-7 orders for oaks from Gillingham, making it possible that the roof still contains some timbers which once came from Gillingham Forest. The distance of the Gloucester house from Gillingham might suggest that there was some particular reason why the Gloucester Dominicans were favoured in this way.

The quest for Gillingham timbers in religious houses need not always take the searcher so far from the forest. In the early fourteenth century the Gillingham parish church of St. Mary the Virgin was rebuilt. Five centuries later the chancel escaped the restoration of the rest of the church, so that today the massive oak timbers of the chancel roof are still to be seen. There is no record of timbers from the forest being used in the fourteenth century building, but it is unlikely that they were sourced from anywhere other than the forest. The same might apply to some other local churches close to the forest, such as St. Peter in Shaftesbury, which still retains its timber roof panels from the sixteenth century; and St. Michael the Archangel in Mere, where extensive medieval timberwork can be found.[19]

In 1267 Henry III gave 12 oaks in Gillingham Forest to the Friars Preachers (Dominicans) to repair the fabric of their church in Gillingham. Nothing further is recorded of a Gillingham friary, and it may be an example of a religious house which was not successful.[20]

Timbers as secular gifts

Other oaks found their way as gifts into the houses and strongholds of secular lords and gentry. In 1253 Richard Earl of Cornwall, the king's brother, was favoured with gifts of oaks for building a castle in his manor at Mere, a mile or so beyond the forest boundary. Richard was a wealthy and powerful figure in English and European politics, and shortly afterwards would be elected King of the Germans, a title that gave him enormous prestige even if it was not universally recognised in Europe. The royal licence to build the castle, and the gifts of timber from Gillingham forest, were likely to have been a gift from the king for his brother's consistent military and financial support.

Mere may have been chosen for the castle site since it lay around half-way between Richard's main power bases in Berkshire and Cornwall, areas in which he held numerous manors, and he may have envisaged it as a point where he could comfortably break his journeys between the two areas. The castle, with its six towers, gatehouse, hall, chapel, and dungeon, was an imposing and dominating feature, overlooking Gillingham and the Blackmore Vale from its mound above the town. In that year Richard had 60 oaks for use in the building of the castle, and had over 130 trees altogether over the period of the mid-thirteenth century. The castle's fortifications never seem to have been called upon in time of anger; the structure was ruinous by the fifteenth century, and today only its mound survives.

Some of the other recipients of Gillingham timbers included:

> *Aymer of Lusignan* or Valence, half-brother of Henry III, 15 oaks (1248), 10 oaks (1258).
> *Nicholas de Molis* or Moels, a commander of royal castles, 20 oaks (1252-3)
> *William of Ralegh*, a powerful judge and ecclesiastic, 20 oaks in 1238 for a house at Warminster.
> *Artaldo of St. Romain*, possibly a foreign favourite of Henry III, 16 arched timbers and 60 beams for building a barn.
> *John le Moyne*, a royal surveyor of estates, 12 oak trees (1269) for 'a certain new house near Shaftesbury, now under construction'.
> *Robert of Mucegros* or Charlton Musgrove (15 oaks between 1236 and 1248)
> *Alan of Plugenet* or Plunkenet in Somerset, 4 oaks (1269), 12 oaks (1297)
> *Margaret Byset*, who is associated with Maiden Bradley, 10 oaks (1238), 10 in 1240 for building a mill.
> *Eustace of Hacce* or Hatch near Tisbury, a powerful knight and royal official, 12 oaks in 1292.

Another powerful and trusted follower of Henry III was John de Lacy, Earl of Lincoln and Constable of Chester. In 1233 the King gave John 10 oaks for his manor at Kingston, now Kingston Lacy, and a further 10 the following year. The building in which they were used may have been a first floor hall house similar to that built at Corfe some decades later. It was also the predecessor of the illustrious and well known mansion of Restoration times.

CHAPTER 8

A WELL-POPULATED FOREST

The word 'assart' comes from the French *essarter*, meaning to grub up woodlands. The concept, if not the word, would have been familiar to the inhabitants of the manor of Gillingham in the twelfth and thirteenth centuries. While it was generally understood that offences against the venison might lead to a heavy fine if detected, the official attitude to offences against the vert was more equivocal. The peasant farmer or cottager who lived on the edge of the forest, or who drove his cows into the forest for pasturage, would note the periodic felling of oak trees for 'official' purposes such as repairs to buildings or royal gifts to outsiders, and might conclude that the preservation of trees as cover for the deer was not perhaps such a high priority after all. Anxious to provide more corn for his family, grazing space for his cows, or firewood for his hearth, he might be tempted to fell some of the forest near his cottage and take in a few square yards of ground. If chance favoured his venture, his enclosure might not be noticed for some time, or at all. Even if the enclosure were detected, the outcome might only be a small fine within the cottager's means.

Reckoning that the enclosure had been worth the payment, the temptation might be to repeat the exercise and gradually enlarge the area around the tenement until it amounted to the size of a small field. It became the practice that such enclosures or 'assarts' from the forest could be made as long as a fine was paid in the courts and a rent thereafter, and that the king's (or queen's) rights, such as the taking of deer, were not infringed. The ground could then be held in copyhold, or by 'copy of court roll.' By the fourteenth century, when royal interest in hunting at Gillingham was declining, fines for new enclosures were becoming a useful source of royal revenue.

The Gillingham manor rent rolls and court rolls have numerous references to assarts and assarting. Some references are to rents paid by tenants for assarts in the forest which they held in addition to their land in the common fields, but there are other entries relating to people whose sole holdings were assarts, with no mention of any holdings in the common fields. Other terms used for enclosures are *encroachments* or *purprestures.* The latter usually refers to grounds for which fines had not been paid, and often included minor infringements for house plots, gardens, hedges, or pathways. The distinction between assarts and purprestures was somewhat legalistic, but the later writer on forests, John Manwood, was able to state that:

> assart is the converting any coverts in the forest into arable; but purpresture is a wrongful encroaching of a new thing either upon the king or a common person … 'tis a purpresture for any man to build a house in the forest, although on his own land … [1]

Assarting gave the forest dweller access to a range of other resources needed for his subsistence. While the felling of mature trees could be undertaken only at some risk or cost, the remains of smaller trees and underwoods might readily find their way into the many humbler uses of the household – repairs to the cottage or outbuildings, firewood, charcoal, thatching spars, hurdles, and many other uses hardly ever recorded in official records.

The full extent of assarting is seen in a rent roll of the manor of Gillingham which dates from around 1300. The wealthier tenants, about a quarter of the population, were holders of virgates, or blocks of strips in the common fields.[2] Further down the social scale were many cottagers with no land at all in the common fields, and dependent on small pieces of ground around their cottages and the grazing of animals on the common wastes and in the forest. Most tenants, rich or poor, appear to have had assarts, which had a lower value than the holdings in the common fields. But some tenants are recording as holding several assarts, which could amount to something more valuable. Henry de Plancis, a wealthy copyholder with two and a half virgates rented at 18s 6d, had several assarts for which he paid 12½d. John the Palmer paid 10s for (probably two) virgates, another 15d for a quarter of a virgate, and a further 13d for assarts.[3]

Some 40 tenants are listed as having tenements (meaning dwellings) and assarts in the forest, but with no mention of virgate land. This might suggest that many tenants were building entirely new cottages or tenements in the newly enclosed grounds. It is easy to envisage how this might arise out of the need to provide living space and cultivable ground for an expanded household which might include newly married children expecting to create a new family. Once a fine had been paid in the manor court for the new tenement and enclosed ground, the new household could enjoy the security of tenure recorded in their copy of the manor court roll.

18 A couple courting in a forest (from the Luttrell Psalter).

The creation of assarts or purprestures might pass us by entirely unnoticed if it were not for the occasional reference to the process amongst the regular, everyday business found in the manor court rolls. In this court roll extract from 1302-3 we read that

> Walter of Wodesend gives to the Lady Queen twelve pennies for a perch and a half of purpresture opposite his door, paying rent one penny a year, by pledge of Richard Wormeswell.[4]

The perch was eventually standardised to a measure of a little over 30 square yards, but in the medieval centuries may have been larger than this in some places. In the c.1300 rental, Walter of Wodesend is recorded as being the holder of a virgate who also had 'sundry assarts' for which he paid 18d. It would seem that this particular purpresture was one of several small forest enclosures which he had in addition to his virgate land in the common field. On the same court roll, the expansion of an existing enclosure might be inferred from this entry:

> Richard Keyne (perhaps Kington) gives to the Lady Queen 3d for an earthwork (or ditch) to be enlarged at Wylrygge, near his enclosure, by pledge of John Barnabie.[5]

All this indicates that the forest was becoming more inhabited by this time, with new fields and cottages gradually being carved out of the oak woodlands. At the time that the huntsmen of John and Henry III were harvesting the deer for the royal table or royal gifts, the deer were finding that the space available for them to freely roam was becoming steadily reduced. The development of the deer park as a space for the fallow deer to find refuge undisturbed was in part a response to the increasing population of the forest, with a growing number of peasant farmers finding the deer a nuisance among their crops and pastures.

The much later map of 1624 shows that most of the new farmland and settlement was around the outer edges of the forest adjoining Gillingham, Shaftesbury, and the Wiltshire boundary, leaving the inner parts clear for the deer park and the forest 'walks' as the effective royal demesne. The division of the forest between walks and enclosures may have been a deliberate policy of wardens and foresters anxious to maintain some spaces clear of human interference.

Motcombe and Enmore Green

The most significant development in the eastern part of the forest was the rise of Motcombe as a separate village. The name, which is not known in Domesday Book (1086), is first recorded in 1244 as *Motecumbe*, and appears in records regularly thereafter.[6] There is no evidence of common fields having existed in or around Motcombe, which helps to confirm that Motcombe had developed entirely as a consequence of assarting and new settlement.

The most likely meaning of the name is 'valley where meetings are held', the meetings referred to being those of the forest attachment and swanimote courts.[7] In Norman times these might have been held at Gillingham, but with the growth of forest population it would have become more convenient to hold the meetings within the forest itself. The meeting place and village grew up around a point not far from where the ancient Mere to Shaftesbury road crosses a small stream called the Cranborne Lake, an accessible venue for many forest dwellers. For some decades the village had no distinctive identity other than as the name of a place in Gillingham forest. In the Lay Subsidy returns of 1327 and 1332 it was still recorded as part of Gillingham.[8]

However, by this time the separate existence of Motcombe was being acknowledged in other ways. In 1319 the parish of Gillingham, which until that time had depended on priests from Shaftesbury Abbey, acquired its own living of a vicarage. The endowment at Gillingham also included 'a house at Motcombe for the priest officiating in that chapel.' The establishment of a chapel at Motcombe with its own priest, although tributary to Gillingham, shows that it was by now developing as a settlement of some size. (Plate 7).[9] At the time of the Lay Subsidy return of 1525 it had become a tithing recorded separately from Gillingham, and may have been so for some considerable time.[10]

A number of names of localities in and around Motcombe first appear at about this time. Some are topographically descriptive names, but in records are linked with rents paid by tenants of the manor of Gillingham. The name *Kyngesettl* (also *Kaingessettle, Kyngessete*) first appears in 1247, and means 'king's seat', perhaps referring to a high place on the hillslopes, or possibly to a hunting lodge of the medieval

19 Short's Green, Motcombe, in the nineteenth century. A William Short is known from 1608, but the locality is a product of forest clearances in medieval times.

forest. In c.1300 rents were paid by Walter son of Istin of Kingsettle, Edith the wife of Istin, and by Simon Kingsettle, all of whom had virgate lands in Gillingham. *Northey* or North Hayes, north of Kingsettle and meaning northern or northernmost enclosure, is a name which occurs in Gillingham court rolls from 1317 onwards. Other names in which the 'hayes' element implies assarting or enclosure from the forest wastes are *Culverhayes* and *Coyteshayes*.[11] These and other names mentioned here can be found on Figure 4 (pp.24-25).

Some place-names first appearing around this time may be linked to a particular person or original occupier of the site. An example may be *Wermeswelle* or Wormswell, between Shaftesbury and Motcombe, in which the reference to a spring or stream is linked with a personal name. The name itself has now disappeared, the place later becoming known as Coward's Shute. Cliff or Cliff House in Motcombe may be the place of origin of William Clive, who was Gillingham's first vicar and is thought to have come from Motcombe. Payne's Place, 'a messuage with appurtenancies in Motecombe … late of William Payn', occurs in a record of 1455. By the following century this had become a substantial residence occupied by a family of some standing. Palmer's, or Palmer's Place, may be linked with a John the Palmer, who c.1300 paid rent for two virgates in the manor and another quarter virgate.[12]

20 *Wormwell in Motcombe, later known as Coward's Shute. A sketch made by Henry Joseph Moule in the 1830s.*

Medieval Motcombe appears to have consisted of clusters of cottages around the church, the later Street, and the later Frog and Shorts Green lanes. Lanes and trackways linked the village with outlying localities, but there was no direct way to Gillingham across the park and forest. Shaftesbury and Mere were more easily accessible by joining the ancient high road or causeway which passed by Motcombe, instead of going through it as it does at the present day.

Another part of the forest which became important at this time for a different reason was Enmore Green. This place, now an outlying part of Shaftesbury, is at the foot of the hill on which the older, main part of the borough was founded (Plate 6). At this point the boundary between the borough of Shaston or Shaftesbury and the forest of Gillingham follows a carefully defined line at the foot of the slope. The boundary must have been established at the time of the foundation of the *burh* of Shaftesbury by Alfred, and medieval perambulations took great care to demarcate it carefully by reference to the boundary points of *Vroggemeare* or Frogmere, a name which suggests a pool or a pit; *Koggemanstone*, a boundary mark or stone; and *Radwelle* or Radwell, a name also relating to a well or pool.

The name first appears in the middle of the thirteenth century, the various early spellings (such as *Enedemere*, *Enmer*, and *Elmer Greene*) appearing to refer to a duck pool or other body of water with a green or open space. A later description refers to 'a pool of water, and divers springs and wells'.[13] The close proximity of Enmore Green to the town strongly suggests that, notwithstanding its location outside the borough jurisdiction, the settlement originated as a suburb of Shaftesbury. Its first inhabitants may have been people from the town content to pay fines to the Gillingham manor court for purprestures and assarts. Lying outside the borough boundary, they would be freed from any feudal obligations required by the town or its abbey.

The various references to wells and springs relate to the boundary between the Upper Greensand, which forms the promontory on which the medieval town was built, and the impermeable underlying Gault Clay, giving rise to a line of springs or wells. These 'three or four large wells' (Hutchins) at Enmore Green took on a particular importance for the town on the clifftop above. Shaftesbury was a flourishing borough which included the largest nunnery in the country, and had grown rich on the gifts of the many pilgrims who flooded into the town attracted by the cult of King Edward the Martyr. However, in choosing a hilltop location, the founders of the town in earlier centuries had made little provision for an adequate water supply. Water had to be carried from sources below the site of the town, and out of this problem there grew an agreement between the borough of Shaftesbury and the manor of Gillingham.

A Gillingham manor court roll of 1527 provides the earliest record of this arrangement, which had existed 'time out of mind'. In return for allowing Shaftesbury townspeople to use the wells of Enmore Green, a ceremony was enacted annually in which Shaftesbury paid homage to Gillingham in an elaborate ritual. The ceremony, known as the 'byzant' ceremony, has been described thus:

> That the Sunday next after Holy Rood day in May, every year, every parish within the borough of Shaston shall come down that same day into Enmore Green, at one of the clock at afternoon, with their minstrels and mirth of game; and in the same green of Enmore, from one of the clock till two of the clock, by the space of one hole lower there they shall dance; and the mayor of Shaston shall see the queen's bailiff have a penny loaf, a gallon of ale, and a calf's head, with a pair of gloves to see the order of the dance that day …

If the ritual and dance did not take place, the queen's bailiff was entitled to stop up the wells. The earliest date at which this ceremony was acted out is not known, but it is thought to have been held more or less continuously between 1364 and 1829, by which time it had become too elaborate and expensive and was discontinued.[14]

Gillingham and Newbury

Further enclosure and new settlement took place along the western side of the forest, bordering on the manor and town of Gillingham. Here the boundary between manor and forest was marked by the water of the Shreen, with the town being clustered around the church close to where the Shreen meets the Stour. By the thirteenth century an outlying part of the town known as Newbury had come into existence within the forest boundary on the higher ground between the Shreen and Lodden. The name, which means 'new manor house' is first recorded in 1295, and may refer to a substantial tenement which existed there.

The 1624 map indicates that early Newbury was a cluster of dwellings around the meeting of the Shaftesbury road with the present-day Hardings Lane, spreading out along the Shaftesbury Road into Ham. From here and northwards, enclosures filled in the gap between the Shreen and the Park Pale, including the capital messuage or tenement of East Haimes, the residence of the fee forester. Much of this land now belongs to Gillingham School. Further north again, cottages and enclosures at Bay and Bowridge Hill (then sometimes spelt as *Porridge* Hill) were the result of assarting and new settlement.

At this period Bay Lane curved round the north of the park towards Donedge Lodge, and there was no through route across the forest towards the Knoyles. *Bengerville* or Benjafield was a substantial forest tenement and is first recorded in 1313. Cottages at Huntingford, close to the northern boundary of the forest, may have originated from medieval assarts.

South and east of Newbury assarts stretched along the area of the south side of the forest boundary to *Sotingestoke, Schetyngstocke,* or Shearstocks. Further south and east again, the forest between Duncliffe Hill and the Shaftesbury road was almost entirely lost to assarts and tenements, extending to Long Cross and Enmore Green. The existence of the Shaftesbury to Sherborne high road through this area may well have made it easily accessible to assarters and new settlers.

By the later thirteenth century, the forest had reached a peak of population which was probably not achieved again until the seventeenth century. The forest laws and the endeavours of foresters, regarders, and verderers had been quite unable to resist the encroaching tide of enclosure and tenements from people able and willing to pay in the manor court for their new land. As a result the open forest had been reduced to half its original size. By contrast, the next centuries were to be a time of little change. The villages, enclosures, and dwellings recorded on the 1624 map would all have been familiar to the forest dwellers of some three centuries earlier.

CHAPTER 9

AFTER KING'S COURT

After 1261 there are no more references to building works at King's Court, and from the time of Edward I the royal visits were much fewer. Neither Edward II nor Edward III are known to have visited Gillingham, preferring the greater comforts of larger palaces such as Clarendon. From this time King's Court became disused and decayed. In 1369, following the death of his queen, Philippa, Edward III ordered the demolition of the palace. In 1403 some of the materials were used to repair the lodge at Merley, the following item being recorded:

> Item to John Towle for digging up the stones of the walls of the old king's court in the park for 19 days, 6s 4d at 4d per day; and to William Thresher for carrying a cart with the stones of the walls of the old court aforesaid, 16s for 16 days at 12d per day.

Hutchins recalls that some of the stone was later used in repairing the Shaftesbury road, and that an attempt to search for the foundations of the palace produced only some small fragments of building stone. In the longer history of Gillingham Forest and Motcombe, the story of King's Court, with its royal splendour, can be seen to have occupied only a brief episode.[1]

From this time the many references to the forest in the royal records such as the Close Rolls come to an end, and with it the detailed picture of forest management and developments. Gillingham is not alone in the drying up of detailed historical evidence of the royal forests, a change which can be partly attributed to the declining interest of the Crown. By 1350 the severe forest laws established by Henry I, designed to preserve the exclusiveness of the forests for royal privilege, had been severely dented by the political and constitutional onslaughts of the barons. Many forests were now more limited in extent than they had been two centuries earlier, while in others Crown control had entirely ceased. In Wiltshire the great forest of Savernake

After King's Court

in the north of the county had been reduced from 98 square miles to a mere 13, while the forest of Chute had almost disappeared. The Dorset forest of Blackmore, bordering Gillingham on its south side and stretching into the centre of the county, had been entirely disafforested. In Somerset the forests of Neroche, Exmoor, and Mendip had been granted away by 1405.[2]

In the remaining forests the royal interest was diminished. The itinerant court of the time of Henry I and John, with its need for a chain of forests and hunting lodges to satisfy the royal leisure time, was now long past. Royal interests became increasingly concentrated on a smaller number of palaces and castles relevant to royal power during the political struggles of the period. While the great palace at Clarendon continued to flourish as the major royal palace of the west of England, the smaller, less accessible outposts such as Gillingham were easily forgotten.

As the royal presence in the forests became more limited, so the machinery of forest administration wound down. Forest eyres became a rarity, so that from 1462 to 1483 the Chief Justice of the southern forests held only four across the whole region. On his accession in 1461 Edward IV attempted to restore the forest system, and sent out orders to elect verderers to nineteen forests south of the Trent. But little could be done to halt the long-term decline of Crown control over the forests.[3]

During the fourteenth and fifteenth centuries the forest of Gillingham may have been less populous than in previous times. Throughout England the early fourteenth century was characterised by poor harvests and economic depression, bringing an end to the growth and prosperity of the previous century. In 1348 the Black Death struck Dorset, and through the later part of 1348 and into 1349 the population of Gillingham and its area may have been as much as halved. The plague was to return again in 1361.[4]

The account of John Nyman

In the Dorset History Centre in Dorchester there is a court roll of the manor of Gillingham which dates from around 1300. On the reverse of one of the sheets some earlier writing has been expunged and replaced by later script. The eye is first attracted by the top of the folio, where there is a sketch of a huntsman and his dog chasing a hare. Further down the manuscript there is a doodle or beginning of a further sketch.[5]

The script casts a beam of light across an otherwise shadowy period in the forest's history. The text is an account produced by one John

21 A huntsman wearing a hen mask pursues a hare in the Forest of Gillingham (from a court roll in the Dorset History Centre)

Nyman, who lived at Motcombe towards the end of the fourteenth century. His account is a list of his expenditure over the year 1378-9. It is not known who John Nyman was, or for what purpose his account was written, but the fact that the account was written at all suggests that he may have had an official post for which written accounts were needed. The mention of a bow and its strings would suggest that he was a forest keeper. The sketch of the huntsman and dog may be a self-portrait of Nyman himself. The inclusion of the place names of *Northey* and (Elena atte) *Clyve* (or Cliff) help us to locate his home to the North Haye area east of the village. We learn that he had a 'croft called Northaye, over against [his] father's house', and close to this the 1624 map marks the field called *Newmans*. In 1327 a Roger le Nieuman had paid 2s 6d in Lay Subsidy, but no Newman is mentioned in the later 1332 assessment.[7]

The account begins in September 1378 by stating that he then owns 34 sheep, including 16 two-year olds, and a *veverhogg*, presumably a type of ram. He also has a part share in a bullock. He records a visit to Shaftesbury market where he sells birds to the value of 10d, then spends the money on the items he needs. His biggest outlay is on beans to sow (4d), but he also buys peppercorns and strings for his bow. Spiritual needs are not overlooked, for he spends 1d on an oblation, most likely for a forbear, and another 1d on a donation.

Around early December he sells a horse and nine lambs, and is given two hedges to ditch and trim in return for keeping all the trimmings to his own use. He afterwards sells the hedges and also 15 pounds of wool. During the spring he hires or leases a croft from Elena Atte Clyve for a year starting at Pentecost. In June he makes another trip to Shaftesbury market, where he buys:

> ...one lamb on the feast of St. John the Baptist 6d; and in ruddle [red ochre for marking sheep] and tar soap 2½d; and in wine and in spice for a medicine 3d; and in strings for my bow, and in one knife 2½d.

By this time the beans which he had sown earlier have finished growing; he notes the yield and the 'turnover', perhaps the surplus ones he can sell. Next comes the haymaking time. He leases the haymaking from a meadow belonging to Elena atte Clyve; his share of the half-day's haymaking is 2½ stacks of hay, all of which he sells to others, but notes that he has still not had the 11d for the half-stack which he sold to Richard Coppedok. Then follows a trip to Hindon market in October, where he buys six sheep. After this some things go badly. One sheep is bitten by a dog, owner unknown, and

> ... moreover Ralph Palmer's dog came, on the Tuesday next after the feast of St. Luke, in one croft called Northay, over against my father's house, and there bit one veverhog, in the second year of the reign of King Richard, and two lambs died of murrain.

The final paragraph is an agreement to lease to William Holme *a certain croft called Fryte*, from that time, which was the Feast of Ss Simon and Jude, 28 October, until Pentecost the following year. Once again, Elena Clyve is part of the picture, since 'he will not trim the hedges without [her] permission.'

Nyman's account shows how he lived and farmed in the countryside around medieval Motcombe, a community created out of the forest assarts and settlements of the previous centuries. He reveals a landscape of small tenant farms, where the farmers are heavily dependent on their sheep and their vegetable plots. Pieces of ground can be leased from or to other tenants. Inhabitants make regular visits to local markets. The market at Shaston or Shaftesbury was confirmed to the town by charter in 1269, although it had been held for many decades previously. In 1392 Richard II confirmed the grants of two markets to Shaftesbury, on Mondays and Saturdays. Nyman's other market was at Hindon, a new town which had been created in 1220 by the Bishop of Winchester in his manor of Knoyle.[6] Hindon's market day was Thursday. Another nearby market which Nyman may have used regularly was at Mere, where the market day was Wednesday.

Some later forest officers

During these centuries the appointment of royal officers to the custody of the forest carried on much as before, despite the general decline in forest administration nationally. In the later fourteenth century the office of warden was held by the Belvale family, lords of Buckhorn Weston. In 1340 John Belevale was granted the bailiwick of the forest of Gillingham which his father Matthew had held. In 1393 a pardon was issued to John Belvale, keeper of the lordship, forest, and park of Gillingham, of all trespasses of vert and venison.[7] In 1402 Belvales's widow, Katherine, found herself in dispute with the abbess of Shaftesbury over her occupation of 40 acres of land outside Shaftesbury, but Katherine Belvale was able to plead that the land lay within the manor of Gillingham and had been granted to her late husband and herself by Edward III.[8]

In 1399 the office of warden was granted for life to a major figure in royal affairs, Sir Hugh Luttrell (c.1364-1428). Sir Hugh's seat of power was Dunster Castle in Somerset, but he had many manors and estates in other counties.[9] From Dunster he gave outstanding military and administrative service to Richard II, Henry IV, and Henry V. Between 1404 and 1415 he was Member of Parliament for Somerset and Devon on several occasions. He took part in Henry V's French campaign, and was consequently appointed seneschal of Normandy. His royal appointments other than Gillingham included from time to time constable of Leeds Castle, constable of Bristol castle, and keeper of Kingswood and Filwood forests.[10]

Luttrell's letters of appointment include the perquisites of the forest to which he was entitled as custodian. These were the pannage, herbage, brushwood, and windfalls; the yearly rents and services from the tenants, described here as 'woodward's tenants', a woodward being an alternative name for a forest keeper; and the profits of the woodcourt or swanimote courts, 'which John Belvale, who held it in grant from the late king, had in virtue of his office; saving to the king sufficient herbage and pannage for the beasts of the wood.'[11]

After King's Court

In one of the few references to timber in the forest at this time, Luttrell in 1402 requested that ancient oaks and rothers from the forest and park should be sold to pay for repairs to lodges and Enclosures within the forest, notably Donedge Lodge and Merley, or Marley, later known as Lawn Farm. The wood sale accounts, compiled by Richard Cressebien and Matthew Vyning, 'are extant, still enclosed in the leather pouch in which they were forwarded to London.'[12] In 1405 Gillingham forest supplied two bucks for Luttrell's Christmas feast at Dunster, a meal which included twelve capons, two bacon pigs, and four bushels of green peas. At this gathering he made gifts to tenants who played to him, and to children who danced for him. In the Dunster accounts for 1420-1, 15d is paid to 'George, my lord's chaplain at Gillingham, for the expenses of my lord there on his return from London.' Luttrell died in 1428, apparently while on a visit to Shaftesbury to see his daughter, Joan, a nun at the abbey. Spices worth 44s. were purchased at Shaftesbury, presumably to embalm his body, which was then carried to Dunster. His last journey may well have taken him through the forest of Gillingham.[13]

22 The tomb of Sir Hugh Luttrell in Dunster church.

In 1439 custody of the forest was given to Sir John Saintlo or St. Loe. His family are associated with Sutton Court near Chew Magna. He was sheriff of Dorset and Somerset in 1436. His grant was much the same as that held by Luttrell, including the keeping of Bristol castle and of the forests of Kingswood and Filwood. The grant was renewed in 1443. Like those to Luttrell, the grants included the dead wood of the forest, the profits of the swanimote courts or woodcourts, and the rents of the woodward's tenants, then worth 43s.[14]

The early Tudor forest

On his accession Henry VII was faced with task of restoring the rule of law to the Crown's forests. An Act of Parliament of 1485 stated that 'Great numbers of persons ... some with painted faces, some with visors and otherwise disguised ... have hunted ... in divers places of this realm ...' and that in future, offenders would be brought before the justices for examination. The forest courts were revived, with swanimotes authorised for many of the southern forests, including Gillingham. The ancient machinery of the forest eyre was set in motion after a long period of disuse. Pleas of the forest of Gillingham were heard at Shaftesbury on 2 September 1489.[15] Those attending included the sheriff of the county, the lieutenant and ranger of the forest, the forester of fee, and other foresters, verderers, regarders, reeves, and tithing men of each of the townships of Gillingham, Motcombe, and Bourton. The business transacted consisted chiefly of assigning the perquisites of oaks, rothers, and deer to the officials, and the registering of claims to timber and pasturage within the forest. The jury presented a list of people who had felled oaks, but in nearly every case could not state the number of trees involved. It was noted that a deer leap in the park pale had been erected without licence since the last eyre, and that it should be removed. It is unlikely that the proceedings amounted to much of a revival of royal authority, and it was the last time that Gillingham was to be troubled with the courts of the forest eyre.[16]

Henry VIII hunted the deer with all the enthusiasm of his Norman and Angevin predecessors, and even created a new royal forest at Nonsuch in Surrey. It did not long survive him. By this time the Crown was assessing the royal forests not as hunting grounds but as sources for timber, especially for ship-building. New officers were appointed as Masters and Surveyors of Woods, one for each side of the Trent, and after 1554 revenues from the forests were managed by 'Surveyors-General of Woods, Forest, Parks, and Chases', responsible to the Exchequer. From this time the authority of the forest justices largely ceased, and forest offenders were prosecuted in the court of Exchequer Chamber instead of the forest eyre.[17]

Nevertheless, from time to time the swanimotes or woodcourts of Gillingham forest could still be aroused into action. In January of 1535 a jury was called upon to hear evidence of a series of deer killings in the forest. A dozen people were involved in the killing of twelve deer over the period November to January. Six deer had been taken in the park, and others at various locations around Motcombe. The ringleader of the group was William Grimston, an inhabitant of Motcombe, a person substantial enough to appear in the returns of the 1545 Lay Subsidy. Two other people were his servants. Two of the others implicated came from Motcombe, and another from Shaftesbury. The jury was told how on November 13 three of the group had

> killed a doe with a halter and a sowre with a greyhound at Wethers and so the same two deer was conveyed to William Weks house and Thomas Weks had the skins of them and at the same Weks house was Grimston and there did break the deer and had thereof with him.

This deer hunt may have been a more sordid, undercover affair than the great *par force* hunts of the Angevin kings, but the group were no less well rehearsed in the techniques required to secure what they needed. A week after this event Grimston went out again with his servants and 'carried at his girdle five halters', clearly anticipating a successful night. It seems that most of the kills were made under cover of darkness, for when the group were apprehended they confessed to killing all the deer 'and to many others that they had killed by night.' They all swore that they had been obeying the orders of Grimston. The record lists no fines nor any indication of sentences.[18]

The episode is a good reflection of the state of the forest by the middle of the sixteenth century. A framework of forest administration remained in being, but it was largely ineffective in safeguarding the rights of the Crown from the population of farmers, cottagers, and outsiders. An astute observer of the time might well look ahead and conclude that a time would soon come when the Crown would want to look for a profitable way to rid itself of the problems of distant forests for which it could find little use.

GLOSSARY OF FOREST TERMS

agister official responsible for supervising commoners' cattle grazing and pannage

alaunt large, powerful mastiff-like dog

assart area of clearance in woodland or waste from which trees have been grubbed up

attachment court a court to receive and enrol the attachments of verderers and foresters of those accused of offending against the vert and venison

bercelet hound which hunts by scent

berner man in charge of hounds

bow and stable method of hunting in which deer are driven towards concealed archers and nets

browsewood cropped shoots and small branches to feed deer

buck male of fallow or roe deer

coppice wood of small trees and underwood grown for periodic cutting

copyhold tenancy by possession of copy of court roll held according to manorial customs and terms

curee ceremony of rewarding hounds on successful completion of hunt

deer leap structure in park pale allowing deer access into deer park but preventing their exit

expediate cut off claws of a dog's forefoot to prevent it from chasing game

fawn young fallow deer

fee forester hereditary forester, often the chief forester

fence month period when forest must be closed and undisturbed to allow fawning

fewterer person in charge of greyhounds

forest eyre highest forest court, held by itinerant justices

grease time period when deer at their best to hunt for food

hart adult red male deer

hind female red deer

lawing see *expediate*

lawn enclosed pasture within forest to provide grazing

lymer scenting hound used in initial quest for deer

mere boundary mark, often delineated by merestones

pale A deer-proof fence of wooden posts and rails forming the boundary of a deer park

pannage pasturage of swine, right to graze swine in woodland *par force (des chiens)* 'by strength of hounds', the ritualised pursuit of the hart

purpresture unauthorised enclosure or building, an encroachment

regarder officer responsible for making inspections of forests to discover trespasses

rothers fallen or dead wood

serjeanty a type of feudal tenure often used for the holdings of royal officers

stag adult red male deer in fourth year

stint limitation on the number of cattle allowed to be kept on common land

swanimote (or swainmote) court to enquire into offences against vert and venison

unharbouring the initial disturbance of the hart to begin its pursuit

venison the beasts of the chase protected by forest law

verderer officer with responsibility of care for vert and venison, appointed by sheriff

vert habitat for deer, green cover, vegetation of forest

walk subdivision of a forest with its own keeper

waste unauthorised felling of wood or underwood

woodcourt In Gillingham, another name for the swanimote and attachment courts

NOTES AND REFERENCES

Abbreviations used in the notes and references:

CCR	Calendar of Close Rolls
CChR	Calendar of Charter Rolls
CIPM	Calendar of Inquisitions Post Mortem
CLR	Calendar of Liberate Rolls
CPR	Calendar of Patent Rolls
CRO	Cumbria Record Office, Whitehaven
CSPD	Calendar of State Papers, Domestic
DHC	Dorset History Centre
GLHS	*Gillingham Local History Society* Journal
PND	*The Place-names of Dorset.*
JRL	John Rylands Library, Manchester
NA	National Archives
PDNHAS	*Proceedings of the Dorset Natural History and Archaeological Society*
SDNQ	*Somerset and Dorset Notes and Queries*
VCH	Victoria County History

Chapter 1 Part of Selwood Forest

[1] Barker, 'Early Medieval Wessex'.
[2] Chandler, *John Leland's Itinerary*, p.431. Thomas Gerard's survey of Dorsetshire, written around 1620, was published in 1732 and attributed to John Coker.
[3] Grant, *Royal Forests of England*, pp.3-14. Watkins A.E. *Aelfric's Colloquy*. Aelfric was a scholar and translator and became abbot of Eynsham in Oxfordshire.
[4] Thorn, Domesday Book, p.56. The Domesday folios record other royal huntsmen in Chippenham, Clarendon, and Savernake forests (Bond, p.122).
[5] Grant, op. cit. Bond, 'Forests … in Medieval Wessex', p.115.
[6] Hawkins, D. (1980), *Cranborne Chase*, pp.175-80. Cranborne Chase was granted away by William Rufus, but came back into royal hands in 1470 (Bond, p.115).
[7] Grant, op. cit.
[8] JRL, Nicholas Papers no. 68. Grant pp.16, 137-9.
[9] Porter, *Gillingham*, pp.23-27.

Notes and References

Chapter 2 Bounds and perambulations

1   Nelson, *Manwood's Forest Laws,* pp.37-40.
2   Bond, 'Forests, Chases, Warrens, and Parks,' p.125.
3   The role of medieval perambulations is discussed in Bond, pp.124-5. The 1225 perambulation of Gillingham is fully described below. Copies of the later ones can be found in Hutchins, p.662, and in JRL Nicholas Papers, nos. 65 (1568), 69 (1638). The 1605 survey is in CRO, Pennington Papers, D/PEN 34. The 1650 'view' of the manor was printed by C.H. Mayo in *SDNQ* for 1923.
4   The *Book of Cerne* is associated with Bishop Aethalwald of Lichfield (818-30). Its only connection with Cerne Abbey is the collection of Cerne documents bound within it.
5   Somerset Record Society, Vol. VIII. pp.123-6, 168.
6   The perambulation is transcribed and translated in Fossett-Lock 'The Cartulary of Cerne Abbey' pp.195-224. See also CPR Hen III 1216-25 pp.567-8).
7   'The hunter's ford', or the ford known to huntsmen, *PND* p.14. The form *Huntleford* or similar is also used in later centuries.
8   Possible derivations of the name Horsington are discussed in *PND* p.27.
9   *PND* p.13.
10 Watts, 'Some Wiltshire Deer Parks', pp.98-102. At Higher Park Farm there was a moated lodge.
11 *PND* pp. 49, 59.
12 *PND* p.50 notes that 'a tradition that the place owes its name to Alfred's resting here on his way to the battle of Edington need not be taken too seriously.'
13 *PND* pp.58, 59.
14 Stour Provost is so-called since soon after the Conquest it was given to the nunnery of St. Leger de Pratellis or Preaux, which founded a cell at Stour.
15 *PND* p.34 gives *Shetenestoke, Schetyngstoke,* and other variants of the name, making it likely that they refer to Shearstocks, half a mile along the Shaftesbury road.
16 The 1300 and 1568 perambulations both started and finished at Barnaby's Bridge.

Chapter 3 The royal visits

1   John's itinerary was established by T.D. Hardy in 1835. Some historians argue that the king was not necessarily present at places where documents were signed on his behalf.
2   Madox, *History and Antiquities of the Exchequer.*

3   See Mortimer, *Angevin England,* pp.17-19, for some of the features which characterised the royal journeys. The quotation is from the *Dialogue of the Exchequer*, written in 1180.
4   Hutchins, p.618.
5   Ibid.
6   Mortimer, ibid. Kanter, 'Peripatetic and Sedentary Kingship'.
7   CCR 1244-51 p.309, CCR 1251-3 pp.290-1, and other entries dated as from Gillingham in CCR, CPR, CChR, CFR. Henry's destinations are itemised in *The Itinerary of Henry III*, compiled by T. Craib (1923), NA.
8   Church, 'Some Aspects of the Royal Itinerary', pp.31-45
9   Church, ibid. Outside the hunting seasons, dogs could be lodged with trusted knights or king's servants around the country and collected as needed.
10 Church, ibid.
11 Chandler, J (2003), *A Higher Reality,* p.65. See also C. Bullock-Davies in *English Historical Review*, 1965.
12 Mortimer, ibid.
13 CCR 1256-9 p.9. CCR 1272-9 pp.219, 262.
14 Colvin, *History of The King's Works,* pp.241-3.

Chapter 4 *King's Court: Palace and park*

1   Hutchins, p.618
2   Wagner, *The Church of St. Mary,* p.8.
3   RCHM, *Inventory of Historical Monuments*, IV, pp.51-2.
4   Some variations in hall types are discussed in Mortimer, pp.9-11. See also Wood, *The English Medieval House.*
5   A full account of the palace at Cranborne can be found in RCHM, *Inventory of Historical Monuments* III, Part 1, pp. 8-11.
6   Wood, ibid. p.72.
7   Wood, p.396.
8   James and Robinson, *Clarendon Palace.*
9   Barnes W.M. (1898) 'Dorset and King John, Part III. Notes on the Pipe Rolls of that Reign.' *PDNHAS* vol.19, pp.65-81.
10 Colvin, *History of the King's Works,* Vol.1 pp. 81-2.
11 Colvin, ibid. The payments to custodians of the more prestigious houses were notably bigger, eg. Feckenham and Windsor 4d per day.
12 Pipe Rolls, Barnes, *PDNHAS* vol.15.
13 CLR 1226-40,1240-5. Colvin, pp.120-30.
14 CLR ibid. In 1319 a vicarage was established at the church in Gillingham, together with the maintenance of a priest in a chapel at Motcombe.
15 CLR 1240-45.
16 CLR 1245-51, CCR 1246-51. Colvin, p.122.

Notes and References

17 CLR 1245-51; CCR 1247-51; CCR 1251-3. Wood, op.cit, p.72.
18 CLR 1245-51; CCR 1251-3; CCR 1252-3.. Hutchins, p.619.
19 CCR 1251-3. The escheator was an officer who had the duty, when a Crown tenant died, of seizing back into the hands of the Crown all his lands held in chief.
20 Hutchins pp.618-9.
21 CCR 1254-6; CCR 1256-9. Hutchins p.619.
22 Hutchins, p.618. CCR 1227-31.
23 Bond, 'Forests, Chases, Warrens, and Parks' p.135 (map). For more detail on all the Dorset deer parks see the gazetteer and descriptions by Cantor and Wilson in *PDNHAS* vols. 83-100.
24 Bond, pp.137, 138 (maps).
25 Birrell, 'Deer and Deer Farming in Medieval England', pp.112-26.
26 Bond, p.140.
27 Cantor and Wilson, 'Medieval Deer Parks of Dorset' Part 5, *PDNHAS* vol. 87, pp.223-33.
28 Baillie-Grohman, *The Master of Game*, p.188.
29 Richardson A. in Liddiard, *The Medieval Park: New Perspectives*, discusses the aesthetic aspects of deer parks in relation to royal residences.

Chapter 5 Forest guardians and justice

1 NA FR C60/18 m10 no.6
2 Grant, pp.88-92. In 1236 the office was divided into separate justiceships for 'this side' (south of) Trent, and 'beyond Trent'.
3 Grant, pp.94-111.
4 The biographical notes in these paragraphs are derived from various sources, including CPR, CCR, and Fine Rolls; standard authorities such as the Victoria County Histories; and several online sources.
5 The De la Linde family of Winterborne Clenston is described in detail in Hutchins vol.1 pp.188-9.
6 Grant, ibid., pp.125-8.
7 CCR 1234-7; CCR 1288-96.
8 Grant, ibid.
9 CCR Edw I 1 (1272-9).
10 Grant, ibid.
11 CIPM 5 Edw II.
12 FR C60/30 m9 1230-1 159. Haime, pp.2-8.
13 Haime, *The Haimes*, explores the members of the Haym family at this time and in subsequent centuries..
14 Haime ibid. For William Bydyk see CIPM, 1427-42; also Mowbray, *History of the Noble House of Stourton*, vol. 1. p.154.
15 NA E101/141/12 and E101/141/13, Certificates of Regarders 1569-1600.

[16] Manwood, *Treatise of Forest Laws.*
[17] Grant R. (1991), *Royal Forests of England*, p.40.
[18] JRL, Nicholas Papers, no.65. See also A. Horsfall in *SDNQ* 33 (1995) p.419.
[19] Ibid.
[20] NA E32/11 54 Hen III (1269)
[21] CPR 1301-7; CPR 1313-17.
[22] CPR 1334-8.

Chapter 6 'The Master of Game'

[1] Berners, *Book of St. Albans,* p.88. The book is attributable to Dame Juliana Berners of Sopwell Priory, near St. Albans. Its essay on hunting is thought to be based on an older treatise written by the French writer Guillaume Twici.
[2] Bond, 'Forests, Chases, Warrens, and Parks', p.125.
[3] Essays *passim* in Liddiard, *The Medieval Park*. Bond, pp.125-6, 142.
[4] Hawkins D (1980) *Cranborne Chase* p.60. Bond, p.126.
[5] Liddiard, ibid. Bond, pp.144-6.
[6] Bond, ibid. NA E32/152/5d.
[7] Birrell, 'Deer and Deer Farming in Medieval England', pp.112-26. Bond, p.128.
[8] Cummins, *The Hound and the Hawk*. Bond, p.126.
[9] Cummins, pp.68-73.
[10] The story is detailed in Thomas Gerard's *Survey of Dorsetshire*, pp.98-9.
[11] *The Master of Game*, written by Edward, Duke of York, is based on an earlier French writing by Gaston de Foix, or Phoebus. Cummins, *The Hound and the Hawk: the Art of Medieval Hunting*, is a modern explanation of medieval hunting practices.
[12] Cummins, Ch.1.
[13] Cummins, ibid.
[14] As described in *Aelfric's Colloquy*. See Chapter 1 of the present work.
[15] Cummins, Ch.3 and subsequent chapters.
[16] Coke, *Debate between the Heralds of England and France.*
[17] Cummins, p.215, based on a reading of a French hunting account, *Le Roman des deduis.*
[18] CCR Edw II 2 1313-18.
[19] Cummins, p.183.
[20] CLR Hen III 1240-45 pp.64, 274. CCR Hen III 1242-7 p. p.437; 1251-3 p.285; 1256-9 pp.9, 251.
[21] Bond, pp.127-8.
[22] The remaining references in this chapter are derived from the Calendars of Close Rolls between 1227 and 1296. It has not been thought necessary to give a specific reference for each item mentioned.

Notes and References

Chapter 7 The green cover

1   Green, *A Landscape Revealed*, portrays in detail the complexity and density of prehistoric settlement on the Dorset chalklands.
2   Birrell, p.117.
3   CPR 1216-25 pp. 360-1. A. Horsfall in *SDNQ* (1995) 33 p.419 Woodcourt of the royal forest, standing orders Eliz.1.
4   Fry, Dorset Inquisitions, no.49.
5   CCR 1227-31.
6   CCR 1237-42.
7   CCR 1247-51.
8   CCR 1247-51.
9   NA SC8/200/9968A, SC8/200/9978B. At this time the church of Gillingham had no living of its own and belonged to Shaftesbury Abbey.
10 Hutchins p.651.
11 NA E32/11 54 Hen III (1269)
12 CCR 1246-51; CCR 1251-3. The uses of all the oak timbers from Gillingham have been tracked in considerable detail by Bill Shreeves in his series of articles 'Where did all the Gillingham Oaks Go ?', *GLHS* 1997-2010. For further detail of the building of the royal castles see Colvin, *The History of the King's Works*.
13 The references in the following paragraphs are all from CCR and can be followed up in the *GLHS* articles.
14 Remaining references in this section from CCR and *GLHS* articles above.
15 See also *VCH Wiltshire* vol. 3 (1956) p.329.
16 See also *VCH Wiltshire* vol. 3 pp.330-1.
17 *VCH Wiltshire* Vol. 3 pp. 295-302.
18 *VCH Hampshire* vol.2 pp.168-72.
19 RCHM Dorset Vol. IV, pp.27-9, 63. See also Wagner, *Church of St. Mary the Virgin*, and Skelton-Wallace, *Selection of Parish Churches*, pp. 48-59.
20 *VCH Dorset* p.92. The 1624 map of the forest marks a *Frarye* on the west bank of the Shreen on land belonging to the rectory of Gillingham church.

Chapter 8 A well populated forest

1   *Manwood's Treatise of the Forest Laws*, ed. Nelson W. (1717), p.312.
2   A virgate was around 30-40 acres, enough to maintain a large, possibly extended family, providing food for the year and a small surplus which could be sold in local markets.
3   DHC D/GIM/M1 Rent roll of the manor of Gillingham c.1300. A transcribed and translated version is in the Gillingham Museum.
4   *SDNQ* XXVIII, 1958-9, no.134, minutes of Gillingham Manor Court 31 Edw 1302-3, translated by Joseph Fowler.

5   Ibid.
6   A selection of early spellings of the name, with sources, is given in *PND*.
7   Ibid.
8   Rumble, *Dorset Lay Subsidy Roll of 1327*. Mills, *Dorset Lay Subsidy Roll of 1332*.
9   Hutchins, p.645.
10 Stoate, *Dorset Tudor Subsidies*.
11 *PND ibid*.
12 *PND* ibid. A *palmer* was a pilgrim who had returned from the Holy Land.
13 *PND* p.48.
14 Hutchins p.629.

Chapter 9 After King's Court

1   Hutchins, p.619.
2   Grant, pp.169-70.
3   Grant, ibid. Bond, p.124.
4   Forrest, 'The Black Death in Dorset: the crisis of 1348-1349', p.3.
5   Fowler, 'A Composite Gillingham Document of the Fourteenth Century', pp.218-20
6   For the background to the new town of Hindon see Beresford, 'The Six new Towns of the Bishop of Winchester', pp.200-3.
7   CPR 1338-40 p.468. CPR 1391-6 p.422. Belvale may have died shortly after this, since two years later a grant for life was made to Roger Crophull of the keepership of the forest and park of Gillingham, as formerly held by John Belvale (CPR 1391-6 p.527).
8   Katherine Belvale was a nurse to Edward III's queen Philippa. The reference to the dispute with the abbess of Shaftesbury is from the online History of Parliament, reference to John Whiting.
9   His family were descended from the Luttrells of Irnham in Lincolnshire. Sir Geoffrey Luttrell (1276-1345) is known as the originator of the famous *Luttrell Psalter*.
10 Kingswood and Filwood are now parts of southern Bristol.
11 CPR 1396-9.
12 CPR 1401-5 p.179 and NA SC8/171/8506. Wood sale account from NA E101/141/11. Leather pouch reference from Cox, pp.330-2.
13 Maxwell Lyte, *History of Dunster*, Part 1, pp.82, 93, 103.
14 CPR 1436-41 p.257; CPR 1441-6 p.142.
15 Grant, pp.181-4.
16 NA DL39/3/2. Cox, pp.130-2.
17 Grant, ibid.
18 NA SC2/170/17.

BIBLIOGRAPHY OF WORKS USED

Baillie-Grohman W.A. and F (1909), *The Master of Game*, by Edward, Second Duke of York.

Barker, Katharine (1984), 'Institution and Landscape in Early Medieval Wessex', *PDNHAS* vol. 106.

Beresford, Maurice W. (1959) 'The Six New Towns of the Bishop of Winchester, 1200-1255', *Medieval Archaeology* No.3

Berners, Dame Juliana (1486), *The Book of St. Albans*, edited with introduction by Blades, William (1901).

Birrell J (1992) 'Deer and Deer Farming in Medieval England', *Agricultural History Review* vol.40.

Bond, James (1994) 'Forests, Chases, Warrens, and Parks in Medieval Wessex', in Aston, Michael, and Lewis, Carenza, eds, *The Medieval Landscape of Wessex*. Oxbow.

Cantor L.M. and Wilson J.D (1961), 'The Medieval Deer Parks of Dorset' Part 1, *PDNHAS* vol. 83.

Cantor L.M. and Wilson J.D (1965), 'The Medieval Deer parks of Dorset' Part 5, *PDNHAS* vol. 87.

Chandler, John (1993), *John Leland's Itinerary: Travels in Tudor England*. Alan Sutton.

Chandler, John (2003), *A Higher Reality: The History of Shaftesbury's Royal Nunnery*. Hobnob Press.

Church, S.D. (2005) 'Some Aspects of the Royal Itinerary in the Twelfth Century', in *Thirteenth century England XI, Proceedings of the Gregynog Conference*.

Colvin, Howard M, ed., (1963) *The History of the King's Works,* Vol.1. HMSO.

Cox, John Charles, (1905) *The Royal Forests of England.*

Cummins, John (1988), *The Hound and the Hawk: the Art of Medieval Hunting*. Weidenfeld & Nicholson.

Forrest, M. (2010), 'The Black Death in Dorset: the crisis of 1348-1349.' *PDNHAS* vol.131, p.3.

Fossett-Lock B. (1908) 'The Cartulary of Cerne Abbey (The Book of Cerne)'. *PDNHAS* vol. 29.

Garmonsway, G.N. ed. (1953), *The Anglo-Saxon Chronicle*, pp. 32-3, 40-1, 149. Dent.

Gerard, Thomas (1980), *Coker's Survey of Dorsetshire* (reprint of 1732 edition, wrongly attributed to John Coker). Dorset Publishing.

Grant, Raymond (1991), *The Royal Forests of England*. Alan Sutton.

Green, Martin (2000), *A Landscape Revealed: 10,000 Years on a Chalkland Farm*. Tempus.

Haime John W. (1970), *The Haimes: A Dorset Family*. Privately published.

Hawkins, Desmond (1980), *Cranborne Chase*. Dovecote Press.
Hutchins, John (1774), The History and Antiquities of the County of Dorset, vol. 3.
Kanter, J.E. (2011) 'Peripatetic and Sedentary Kingship: The Itineraries of John and Henry III', in Burton J. et al. (ed), *Thirteenth Century England*. Boydell Press.
Liddiard, R. (2007) *The Medieval Park: New Perspectives*. Windgather Press.
Madox, Thomas (1711), *The History and Antiquities of the Exchequer of the Kings of England*.
Maxwell Lyte, Sir Henry C. (1909), *A History of Dunster and of the families of Mohun and Luttrell*. Part 1.
Mills A.D. ed. (1971), *The Dorset Lay Subsidy Roll of 1332*. Dorset Record Society Vol. 4.
Mills A.D. (1989) *The Place-Names of Dorset*. English Place-Name Society, Part III.
Mortimer, Richard (1994), *Angevin England, 1154-1258*. Blackwell.
Mowbray, Charles B.J. (1899), *History of the Noble House of Stourton in the County of Wiltshire*.
Nelson, William (1717), *Manwood's Treatise of the Forest Laws,* 4th edition.
Porter, John (2011), *Gillingham: The Making of a Dorset Town*. Hobnob Press.
Royal Commission on Historical Monuments (1972), *Inventory of Historical Monuments in the County of Dorset*. Vol. III, Central Dorset, Part 1.
Royal Commission on Historical Monuments (1972), *Inventory of Historical Monuments in the County of Dorset*. Volume IV, North.
Rumble A.R. ed. (1980), *The Dorset Lay Subsidy Roll of 1327*. Dorset Record Society Vol. 6.
Shreeves, William (1997-2010) series of articles entitled 'Where did all the Gillingham Oaks Go ?', *GLHS*.
Skelton-Wallace, Jack (2004), *A Selection of Parish Churches in and around …the Blackmore Vale*.
Stoate T.L. (1982), Dorset Tudor Subsidies granted in 1532, 1543, 1593. Bristol.
Thorn, Caroline and Frank (1983), *Domesday Book for Dorset*. Phillimore.
VCH of Dorset (1908) Vol.2 ed. Page, W.
VCH of Hampshire (1903) Vol. 2 ed. Doubleday, A
VCH of Wiltshire (1956) Vol. 3 ed. Pugh, R.B. and Crittall, E.
Wagner A.F.H.V (1956) *The Church of St. Mary the Virgin, Gillingham*.
Watkins, A.E. (undated) *Aelfric's Colloquy,* Kent Archaeology Paper no. 016.
Watts, K. (1998), 'Some Wiltshire Deer Parks', *Wiltshire Archaeological and Natural History Magazine,* vol. 91.
Wood, Margaret (1965) *The English Medieval House*. Harper Collins.

INDEX

Admundestrop, 63
Aelfric, 7
Agisters, 8, 46, 98
Alaunt, 59, 98
Alcester, 15
Alemore, 71
Alfred, king, 7, 60, 87
Alfred's Tower, 6, 10
Angouleme, see Engolisma
Anselm, Archbishop, 19
Articles of Regard, 45
Assarts, 45, 81-3, 98
Attachment, 47, 51, 84, 98
Aundeley, Maurice, of, 43
Aundeley, Robert of, 31-2, 43, 63
Avon, river, 9

Barnabie, John, 83
Barnabie's bridge, 16
Barrow gate, 38, 48
Barrow Street and Lane, 13
Basingstoke, manor, 65
Bath, 8, 21, 66
Bay and bridge, 16, 8-98
Bec, abbey, 15
Belvale, John, Matthew, family, 94
Bengervill, Benjafield, 16, 89
Bere, forest, 8, 43
Bere Regis, 17
Berner, 59, 98
Biggleswade, 19
Biset, Byset, John, 72; Margaret, 80
Bitene, 15
Black Death, 91
Black Ven, 16
Blackmore, forest, 8, 44, 56, 60, 90; Vale, 10, 67
Blandford, manor, 40
Blind Lake, 15, 71
Book of St. Albans, 55
Bounds of forests, 11-16
Bourdeaux, 65
Bourton, 96; Robert of, 33
Bow and stable hunting, 38, 60-61, 98
Bowbearer, 50
Bowridge, bridge and hill, 16, 88
Breamore, priory, 77, 78
Bridewell Lane, 14

Bridport, manor, 34
Bristol, castle, 94, 95
Broomfield, 44
Bryanston, 40
Bugley, William of, 49
Bydyk family, 49
Byzant ceremony, 88

Caen, Roger of, 74
Camel, 74, 75
Candeur, Henry of, 63
Castle Hill, Shaftesbury, 10
Cerne, abbey, 7; *Book of* and Cartulary, 12;
Cerne, Henry of, 12;
Chancery, department of, 20
Charlton Musgrove, 36
Charterhouse, 17
Chew Magna, 95
Chief Justice of the forest, 8
Chiens de courant, 59
Chute, forest, 90
Clarendon, Assize, 18, 7; forest, palace and park, 9, 17, 19, 23, 27, 29-30, 38, 42, 74
Clear Walk, 10, 50
Cliffe, Cliffe House, 10, 86, 92
Clifford, Roger of, 42
Clive, Clyve, Elena atte, 92-3; William, 86
Close Rolls, 42, 72, 73
Cockemanstone, and Cokeman family, 15, 87
Colesbrook, 16
Commissions of enquiry, 54
Coneygarth, 39
Coppedok, Richard, 93
Coppicing, 68-9
Copyholds, 81, 98
Corfe, castle, 19, 33, 35, 43, 44, 74, 75
Cornwall, Richard, earl of, 14, 64, 79
Corsham, 64
Court, itinerant nature, 18-21, 23
Cowards Shute, 15, 86
Cowridge, 14
Coyteshayes, 85
Cranborne, palace, 17, 19, 27, 35; Chase, 8, 56
Cranborne Lake, 53, 84
Cranford, Nicholas of, 71

109

Cresseben, Richard, 33, 95
Culverhayes, 85
Cusin, William, 15

Dean, forest of, 42
Deer, types of, 55-56; deer leap, 37, 98
Deer parks, Dorset and Gillingham, 36-9
Devizes, 17
Dialogues of Aelfric, 7
Domesday Book, 7
Donedge Lodge, 38, 89, 94
Donhead, 36, 54
Dorchester, 17, 74, 75
Drumare, Luke of, 65
Dublin, 66
Dugdale, William, 40
Duncliffe, Dunkweie, 10, 14, 15, 60, 71, 89
Dunster Castle, 94-5

East gate, 38
East Haimes, 48, 49, 53, 88
East Knoyle, 36
East Stour, 15
Edward, Duke of York, 39
Edward I, king, 9, 12, 19, 23, 65, 77, 90; prince, 34, 42
Edward II, king, 23, 49, 90
Edward III, king, 23, 54, 90
Edward IV, king, 91
Edward, huntsman, 7, 48
Edward the Confessor, king, 7, 34-5
Edward the Martyr, king, 34-5, 87
Ela, countess of Salisbury, 64
Eleanor (of Provence), queen, 44, 78
Elizabeth I, queen, 12, 52
Elstow, abbess of, 66
Encroachments, 45, 82
Engolisma, Iterius of, 65
Enmore Green, 10, 87-8
Exchequer, department of, 20
Exeter, 17
Exmoor, forest, 91
Eynesham, abbey, 7

Falconry, 61-2
Falemore, 71
Farming and crops, 93
Farnham, 17
Fawning, 57
Feasts, provisions for, 21-3

Fecamp, Audree de, 66
Feckenham, 31
Fee forester, 7, 46-50
Fence month, 51, 57, 98
Fernbroc, Fernbrook, 26
Ferngore, Frengore, 14
Ferrers, Agnes, 66, 77
Fewterer, 59, 98
Filwood, forest, 94, 95
Fishponds, 39
Forest, Charter of the, 8, 53
Forest Deer, 5
Forest eyre, 40, 48, 53-54, 72, 98
Forest Farm, 5
Forest laws, 8
Forest Lodge, 5
Foresters, 50, 57, 98
Forests, origins and extent, 5, 67-68; decline, 90-91, 96-7
Fremantle, Fremanton, 17, 19, 33
Frog Lane, 86
Frogmere, 15, 87

Gault Clay, 87
Geldenemore, 71
Gerard, Thomas, 6
Gifts, of deer, 64-66; of timber, 76-80
Gillingham, church, 71-2, 78, 84; friary, 79; town, 5, 8, 9, 88, 96
Gloucester, Dominican friary, 77, 78
Goce, *see* Joce
Godmanneston, Ralph of, 33
Grazing, *see* pasturage
Grease, time of, 57, 98
Guildford, 19, 38
Gutch Pool, 14
Greensand, 9, 14, 67, 87
Grenville, Adam de, 73
Grimston, William, 96-7

Hacce, Hache, Hatch, 65; Eustace of, 80
Haime, Haym family, 49
Ham, 13, 16, 70, 88
Hardings Lane, 88
Hart, status and hunting, 57-60
Haselholt, 13
Hawking, *see* falconry
Hayward, John, 49
Henry I, king, 19, 90
Henry II, king, 8, 12, 18, 74

Index

Henry III, king, 12, 19, 29, 30-31, 33, 35, 39, 42, 45, 58, 74
Henry VII, king, 95-6
Henry VIII, king, 6, 96
Herbage, 41
Heybote, 48
Heyning, 57
Hindon, 93
History of Dorset, *see* Hutchins, John
Holme, William, 93
Hore Apeldure, 13, 14
Hounds, types of, 21, 59
Hunting, Saxon period, 7; hunting parties, 20-21; practices, 58-64; for larders, 62-4
Huntingford, 5, 13, 89
Husbote, 48
Hutchins, John, 26-7

Ilchester, 53
Insula or Lisle, Brian de, 12, 22, 40, 42
Iwerne Wood, 56

Joce, John, and family, 48-9
John, king, 5, 17-23, 26, 27, 28, 30-1, 35, 39, 58, 74, 76
Justices of forests, 8, 41

Keepers, *see* foresters
Keyne, Richard, 83
Kimmeridge Clay, 9
King's Bridge, 16
King's Court 5, 14, 16, 43, 48, 53, 60, 68, 71, 73; as royal venue, 17-23; building and development, 26-36; decay, 90
Kingsclere, 17
Kingsettle, 9, 10, 14, 67, 85; Walter, Istin, and Simon of, 85
Kingston Lacy, 56, 80
Kingswood, forest, 94, 95
Kington Magna, 36
Knapp Hill, 14
Knoyle, manor, 71, 93
Koggemanstone, *see* Cockemanstone
Kurhigge, *see* Cowridge

Lacock Abbey, 64
Lacy, John de, 80
Lancelevere, John of, 12
Langley, Geoffrey de, 43
Laund, Lawn Walk 10, 50
Lawing, of dogs, 52

Lay subsidies, 84
Leeds, castle, 94
Legh, Leighe, La, 13;
Leland, John, 6
Leveresmore, 71
Liberate Rolls, 72, 73,
Lincoln, cathedral, 19
Linde, John de la, 44
Little Down, 9, 15
Lodbourn bridge, 16
Lodden, bridge, 16; river, 9, 10, 14, 16, 21, 38, 88
London 17, 18; Tower of, 35
Long Cross, 89
Longespee, Nicholas, 64; Stephen, 64; William, 22, 64
Lovel, John, 62, 63; William, 63
Lower Park, 38
Ludgershall, 38, 42
Lusignan, Aymer of, 80
Luttrell, Joan, 95; Sir Hugh, 94-5
Lymer, 59, 98
Lynde, Thomas de la 58

Madjeston, 13, 44, 70
Maiden Bradley, priory, 77, 78
Malmesbury, William of, 7
Manwood, John, 11, 50, 82
Mares, Robert des, 63
Margaret, queen, 44, 49
Marie, abbess of Shaftesbury, 22, 76
Markets, 93
Marlborough, 17, 19, 42
Marleghe, Merley, 70, 72, 94
Marnhull, 8
Marshall, William, 20, 22
Master of Game, The, 38, 58-60
Medieval halls, 28-9
Melbury Abbas, 76
Melksham, 35
Mendip, forest, 91
Mere, and manor, 13, 14, 70; castle, 64, 79; church, 79; deer park, 14, 36; market, 93
Mere Water and Mereford, 16
Milton (on Stour), 70
Milton, abbey, 77, 78
Moels, Molis, Nicholas of, 65, 80
Monemue, Monmouth, John de 42, 69
Montacute, priory of, 13, 16, 70, 77, 78
Montacute, William, 22

111

Monte Sorrello, William of, 34
Montfort, Simon de 42
Morville, William of, 12
Moyne, John le, 80
Motcombe, 5, 10, 14, 38, 53, 59, 67, 84-7, 91, 96; chapel, 85; Park, 10;
Mucegros, Musgrove, Robert de, 66, 80

Neroche, forest, 91
Nettyates Lane, 14
Neville, Alan de 12, 42; Hugh de, 12, 22, 31, 42
New Forest, 8, 23
Newbury, 9, 88
Newgate Bushes, 14
Newman, Roger le, 92
Newton, William, 27
Nonsuch, 96
Northey, North Hayes, 85, 92, 93
Nottingham, sheriff of, 33
Nyman, John, 91-2

Pale Mead, 38
Palmer, John the, 82, 86; Ralph, 93
Pannage, 41, 46, 98
Par force hunting, 58-60, 99
Park Farm, 38
Park keeper, 37
Park Pale, 5, 88
Parker, Parcere, John and Richard, 50
Passelewe, Robert, 43
Pasturage, 69-70
Patent Rolls, 42
Payne's Place, 86
Perambulations, 11-13
Pershore, abbot, 33, 35
Philippa, queen, 54
Pierston, 70
Pimperlegh, Pimperleaze, 13
Pipe Rolls, 30, 33, 35
Plancis, Henry de, 82
Plucknett, Preston, 44
Plugenet, Plukenet, Alan of, 44, 66, 80
Poitevins, 42-3
Pollarding, 68-9
Pomweyford, Pulverford, 16
Porridge, Powridge, *see* Bowridge
Portway, 71
Port Regis, 10
Powerstock, forest, 8, 17, 31, 40, 42, 44
Preaux, abbey, 15

Purprestures, 81, 99
Pyle Cross, 15

Queen's green, 71
Queen's House, 71
Queen's Oak, 13

Rabayn, Elias of, 33, 43, 65
Radwell, 15, 87
Ralegh, William of, 65, 80
Ranger, 50
Regarders, 45, 99
Rivallis, Rivaulx, Peter des, 42
Roads, 9, 15
Roches, Peter des
Rothers, 69, 99
Royal Commission on Historical Monuments, 27

St. Eustace, 34
St. Loe, Saintlo, Sir John, 95
St. Romain, Artaldo of, 80
St. Rumbold, 54
Salisbury, 9; bishops, 64, 74; castle, *see* Sarum, Old; cathedral, 22, 77, 78; Friars minor, 77
Sarum, New, *see* Salisbury
Sarum, Old, and castle, 22, 74
Savernake, forest, 90
Scammell, Walter, 70
Sedgehill, 14
Selwood, 6, 10, 67
Selwoodshire, 6
Semley, 14
Serjeanty, 47, 99
Sete, 16
Shaftesbury, 6, 9, 15, 16, 21, 59, 62, 70, 87, 96; market, 93; St. Peter church, 79
Shaftesbury, abbey and abbesses, 15, 22-3, 35, 66, 73, 76-7, 85, 95
Shearstocks, 89
Sherborne, 17, 19, 53; abbey, 9; castle, 9, 33, 43, 73, 74
Sherborne Causeway, 15, 71
Sheriff of Somerset and Dorset, 33, 41-2, 45, 46
Shorts Green, 86
Shreen, river, 8, 9, 13, 16, 88
Somerton, 34, 35, 56
Soulescombe, 14
Southampton, 8

112

Index

Stalbridge, 36
Stephen, king, 19
Stoke Trister, 36
Stour, river, 9, 13, 16, 88
Stour Provost, 8, 15, 54
Stourhead, 10
Stourton, 36, 49
Sturminster Marshall, 21
Sturminster Newton, 8, 17
Sutton Court, 95
Sutton Waldron, 49
Swanimote, 51-52, 84, 96, 99
Swyre, manor, 44

Tamworth, 42
Tarrant Crawford, 77
Tarrant Gunville, 36
Taunton, 19; Guy of, 70
Thorngrove, 53, 71
Thresher, William, 90
Timber gifts, 76-80; rights, 22; timber for building, 72-76; transport, 76
Todber, 8
Towle, John, 90
Treatise of the Forest Laws, 11, 50
Turners Stile, 38
Unharbouring of quarry, 59, 99
Unmaking of quarry, 60

Valence, Aymer de, 64
Venison, beasts of, 55, 98
Verderers, 8, 44, 99
Vernun, John of, 73, 77
Vert, 47, 68, 99
Vroggemeare, *see* Frogmere
Vyning, Matthew, 95

Walerand, Robert, 42

Walks in Gillingham forest, 10
Wardens, 40-4, 94-5
Wardrobe, department of, 20
Warminster, 6
Warren, 39; beasts of, 55-6
Waterloo Farm, 38
Weeks, Thomas and William, 97
West gate, 38
Westminster, 18, 19; abbey, 35
Wethers, *see* Withies
White hart, symbolism, 58
White Hart, forest of, *see* Blackmore, forest
White Hill, 13
Wickhampton, Robert, 54
William II, king, 18
Wilton, 53; abbess and abbey, 14, 54; Dominican priory, 77, 78
Winchester, 17, 22, 23, 64, 65, 71
Windsor, 17, 22, 35;
Windyridge Farm, 48
Winterborne, John of, 48; Walter of, 70
Winterborne Clenston, 44
Withies, 14, 97
Woodcourt, 51, 99
Wood reeve, 50
Woods End, Walk, 10, 50; Walter of, 83
Woodstock, 33; Assize of, 8
Wormwell, Wearmwell, 15, 86; Richard, 83
Writtle, 27
Wyke, 70; John of, 71; Robert of, 33
Wylyrigge, 83
Wyndlam, 70
Wyntershulle, William of, 70

Zeals, 36